Disease Investigation-Third Edition: A workbook for an introduction to epidemiology

Kathleen O'Rourke, RN, MPH, PhD
Professor
Department of Epidemiology
University of South Florida
Tampa, Florida

Jessica Berumen, MPH, CPH
Health Science Specialist
Tampa, Florida

About the authors:

Dr. Kathleen O'Rourke is a professor of epidemiology at the University of South Florida with a specialization in Maternal Child Health research. She has conducted research in local and international settings, having received a Fulbright Award to study the impact of traditional birth attendant training programs in Guatemala. Dr. O'Rourke currently serves as the chair of the Epidemiology and Biostatistics department at the University of South Florida. She has a strong interest in student education.

Jessica Berumen has an MPH in Public Health Practice and a Bachelor's degree in Education. Ms. Berumen is employed by the VA working on a national evidence-based research project studying Traumatic Brain Injury in veterans. Her focus is on putting into practice the fundamentals and foundations of public health study.

DLK PUBLISHING
www.dlkpublishing.com
Edgewater, Florida

PO Box 434 Edgewater, FL 32132
United States of America
Printed in USA

ISBN 978-0-9885144-0-9

Table of Contents

Introduction

This workbook serves as a resource to help you understand the materials presented in Disease Investigation: Introduction to Epidemiology. It includes required and optional study assignments, background material, and exercises to help you improve your epidemiological skills.

If you complete the exercises in the workbook, you will increase your understanding of epidemiology. Thus, the workbook can also serve as a guide in studying for exams. While answers are provided at the end of the workbook for some activities, it is to your benefit to work out the problems prior to reviewing the answers. This workbook is meant to be used in conjunction with Canvas. Be sure to familiarize yourself with Canvas, and its different features. Specific instructions for activities will be provided in the syllabus, and you should refer to the syllabus if you have any questions.

Some students are concerned that Epidemiology could be confusing. It's true this field of study does require new skills; however it is based upon logical thinking. If you take your time and work out the exercises as well as participate in class assignments, you will find yourself mastering the subject matter before you know it.

Module 1: Past and current epidemics: Implications for health

This module provides students with an understanding of the ways in which epidemiology is applied to current problems. A number of important terms are presented, and students will gain an appreciation for some of the ethical concerns in epidemiologic research.

Epidemiology can be used to answer a number of health questions. People often first think of epidemiologists as dealing with infectious diseases... OK, just after they thought it had something to do with skin. But epidemiology is concerned with many more issues, those that relate to the health of the population. In 1948, the World Health Organization described health as "a state of complete physical, mental and social well-being and not merely the absence of disease or infirmity." This definition has not been amended since then.[i]

Since epidemiology is the study of health, epidemiologists are also concerned about issues related to quality of life, mental well-being, and social issues. Health is also of great interest to the population and if you just look at a daily paper or a news feed on the internet you will find many articles related to health and disease. Take a look at the Yahoo, CNN, and MSN home pages and see what articles you can find relevant to health. List at least three topics below:

1.

2.

3.

[1] Preamble to the Constitution of the World Health Organization as adopted by the International Health Conference, New York, 19-22 June, 1946; signed on 22 July 1946 by the representatives of 61 States (Official Records of the World Health Organization, no. 2, p. 100) and entered into force on 7 April 1948.

One challenge in epidemiology is that a single study does not prove anything conclusively. It requires multiple studies to understand an issue, and sometimes studies show conflicting results. But news organizations want "hot off the press" releases so at times they present the results of one study as fact, only to have a different study show an opposite finding in a few years. The reason for these differences might be differences in the study design, different subjects, or a different time period. As students in epidemiology you will learn to understand the reasons for this and to be able to evaluate research studies. In this class, you will have the opportunity not only to read about epidemiology, but also to practice designing research studies. You will be better able to understand stories about health that you read in the popular press. Furthermore, you can impress your friends and families at parties with this knowledge.

This first module presents a crossword puzzle which will allow you to check your mastery of important epidemiologic terms and knowledge of historic figures in the field of study. It's followed by a description of three important pandemics; the Black Death of 1347-50, Influenza of 1918, and the 2012 Ebola Epidemic. At the end of this descriptive section, there is a list of suggested videos which will help you in completing the brochure assignment.

Crossword Puzzle

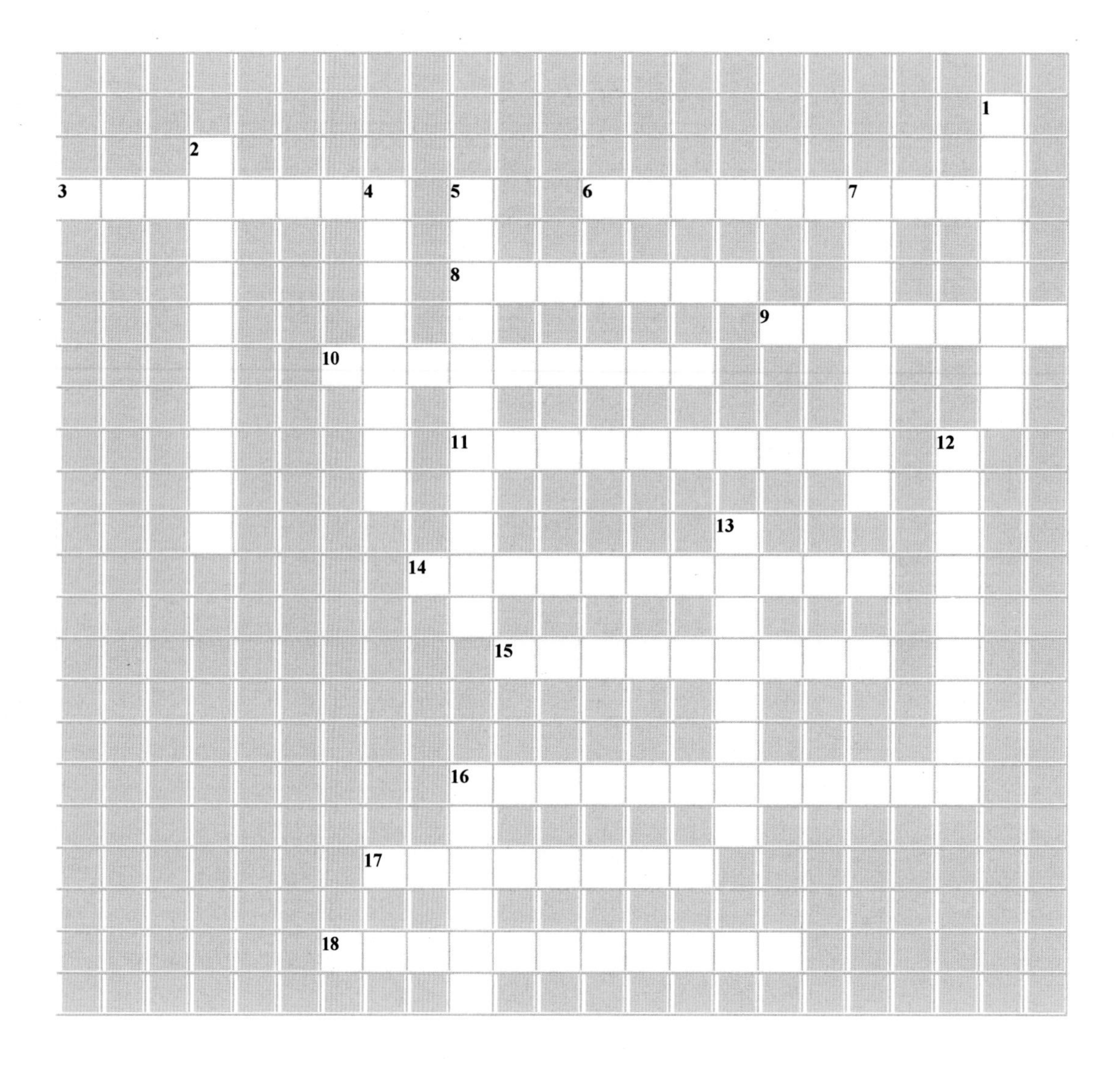

Down

1. Epidemic on a grand scale that crosses international borders and affects a large amount of people
2. Author of the Bills of Mortality
4. The occurrence of an illness/disease in a certain time/place that is clearly in excess of normal expectancy
5. Ancient Greek who contributed to the field of epidemiology
7. Type of intervention that occurs in late pathogenesis stage
12. Illness due to a specific disease/health condition
13. Branch of epidemiology that deals with hypothesis testing
16. Norms or conduct that distinguish between acceptable/unacceptable behavior

Across

3. Contact with disease causing factor or the amount of factor that impinges upon a group or individuals
6. All the inhabitants of a given country or area considered together
8. Type of intervention that occurs before pathogenesis occurs
9. All the possible results that may stem from exposure to a causal factor
10. Type of intervention that occurs in early pathogenesis stage
11. Exposure that is associated with a disease morbidity, mortality or health adverse health outcome
14. Any factor that brings about change in a health condition
15. Causes of death
16. The study of the distribution of disease, determinants and deterrents in human populations
17. Person who demonstrated cholera could be spread by water
18. Branch of epidemiology concerning characterization of person, place and time variables

An overview of three pandemics: The Black Death of 1347-50, Influenza of 1918, and Ebola 2012

Introduction

Pandemics refer to epidemics that spread worldwide. This section provides information on two pandemics that had a marked experience in history.

THE BLACK DEATH

The Black Death was a major pandemic that occurred in the middle of the 14th century, spreading throughout Europe, Asia, and the Middle East. Approximately twenty million Europeans (about 25% of the population) died from this epidemic in six years. In fact, in urban areas that were particularly hard hit, it is estimated that nearly 50% of people died.

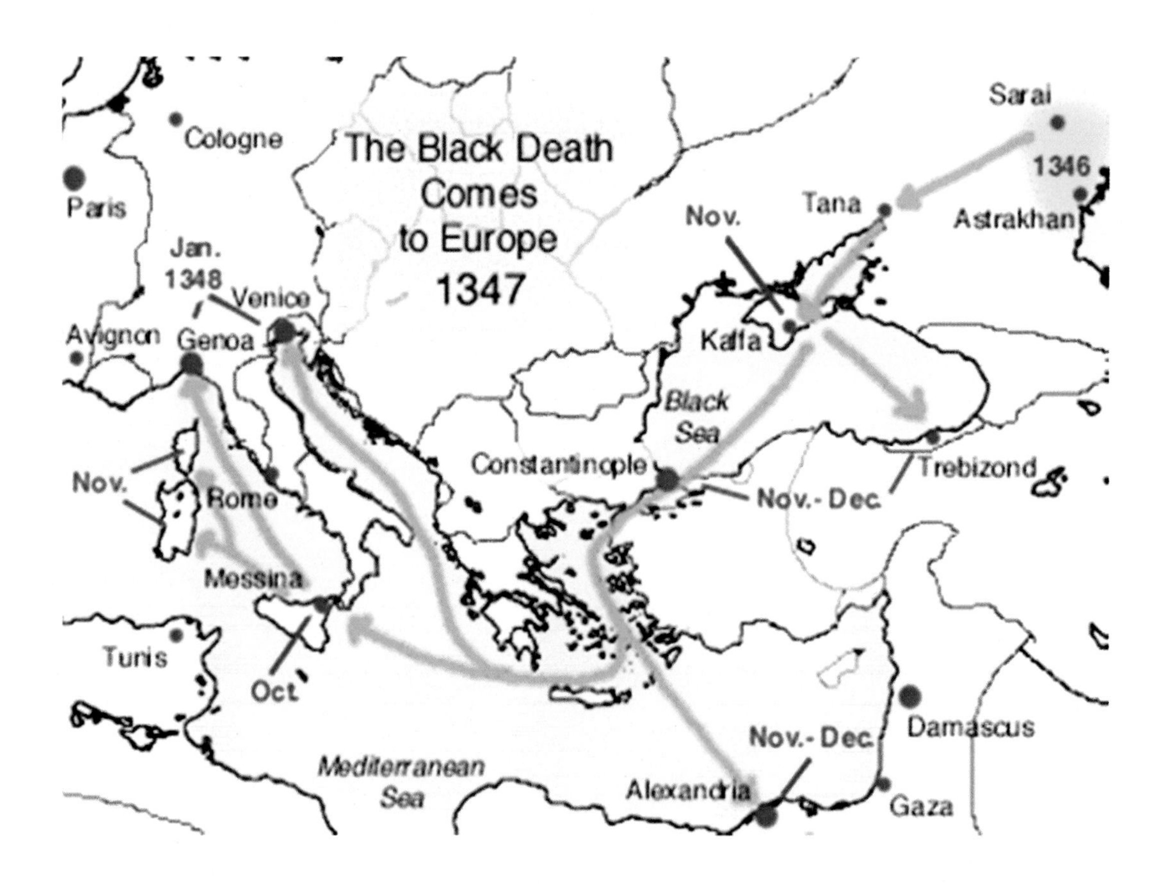

Figure 1. Map of the spread of the Black Death

Figure 2: Typical attire of the medical workers. The beak is a primitive gas mask, stuffed with substances (such as spices and herbs) thought to ward off the plague. "Doctor Schnabel von Rom" (English: "Doctor Beak of Rome") engraving by Paul Fürst (after J Columbina).

It was believed the Black Death arrived in England through Weymouth on June 25, 1348. According to the Grey Friars Chronicle: "In this year 1348 in Melcombe, in the county of Dorset, a little before the feast of St. John the Baptist, two ships, one of them from Bristol came alongside. One of the sailors had brought with him from Gascony the seeds of the terrible pestilence and through him the men of that town of Melcombe were the first in England to be infected." It is thought that the ships carried infected rats and/or fleas. In fact, some ships were found grounded on shorelines, with no one aboard remaining alive. Doctors dressed in strange attire to protect themselves from the illness. See figure 2. Imagine if your friends and relatives started dying and you had no idea what the cause was. Think about losing 1 out of every 4 people close to you. What would you do in this situation? There were no known treatments although many things were tried. Some suggested special diets, bloodletting, and different sleeping positions.

According to the University of Paris Medical Faculty, 1348, "No poultry should be eaten, no waterfowl, no pig, no old beef, altogether no fat meat...It is injurious to sleep during the daytime...Fish should not be eaten, too much exercise may be injurious...and nothing should be cooked in rainwater. Olive oil with food is deadly. Bathing is dangerous." Rich people tried taking medicines made of gold and pearls. Sound was also used and churches rang bells to drive the plague away. There were also many charms and spells that people purchased to fight the plague. People were just desperate and they would try anything in an attempt to find a cure. But in reality nothing worked. Two effective measures though were quarantine and isolation. Although people still died, there were fewer deaths in cities where these quarantines were enforced. Interestingly, Pope Clement VI, living at Avignon, believed the plague was due to poor air quality and so he sat near fires to keep the air pure. In reality, the fires kept the rats away and the bacillus was destroyed by heat so this actually worked but for a different reason.

Here are some quotes from people living at that time:

The victims "ate lunch with their friend and dinner with their ancestors in paradise." **Boccaccio**

"It was dark before I could get home, and so land at Churchyard stairs, where to my great trouble I met a dead corps of the plague in the narrow ally just bringing down a little pair of stairs." -**S. Pepys**

"Neither physicians nor medicines were effective. Whether because these illnesses were previously unknown or because physicians had not previously studied them, there seemed to be no cure. There was such a fear that no one seemed to know what to do. When it took hold in a house it often happened that no one remained who had not died. And it was not just that men and women died, but even sentient animals died. Dogs, cats, chickens, oxen, donkeys sheep showed the same symptoms and died of the same disease. And almost none, or very few, who showed these symptoms, were cured. The symptoms were the following: a bubo in the groin, where the thigh meets the trunk; or a small swelling under the armpit; sudden fever; spitting blood and saliva (and no one who spit blood survived it). It was such a frightful thing that when it got into a house, as was said, no one remained. Frightened people abandoned the house and fled to another." -**Marchione di Coppo Stefani**

"It struck me very deep this afternoon going with a hackney coach from my Lord Treasurer's down Holborne, the coachman I found to drive easily and easily, at last stood still, and came down hardly able to stand, and told me that he was suddenly stuck very sick, and almost blind, he could not see. So I 'light and went into another coach with a sad heart for the poor man and trouble for myself lest he should have been struck with the plague, being at the end of town that I took him up; But god have mercy upon us all!" **-S. Pepys**

"Realizing what a deadly disaster had come to them the people quickly drove the Italians from their city. However, the disease remained, and soon death was everywhere. Fathers abandoned their sick sons. Lawyers refused to come and make out wills for the dying. Friars and nuns were left to care for the sick, and monasteries and convents were soon deserted, as they were stricken, too. Bodies were left in empty houses, and there was no one to give them a Christian burial." **Unknown**

"How many valiant men, how many fair ladies, breakfast with their kinfolk and the same night supped with their ancestors in the next world! The condition of the people was pitiable to behold. They sickened by the thousands daily, and died unattended and without help. Many died in the open street, others dying in their houses, made it known by the stench of their rotting bodies. Consecrated churchyards did not suffice for the burial of the vast multitude of bodies, which were heaped by the hundreds in vast trenches, like goods in a ships hold and covered with a little earth." -**Giovanni Boccaccio**

What caused the Black Death?

Scientists and historians generally believe the Black Death was an incidence of bubonic plague. This epidemic was caused by the bacterium *Yersinia pestis* which is spread by fleas (Figure 3) carried by rodents. Bubonic plague is transmitted through the bite of an infected flea or exposure to infected material through a break in the skin. Although it is not contagious from one person to another, it can spread rapidly in areas with poor sanitation.

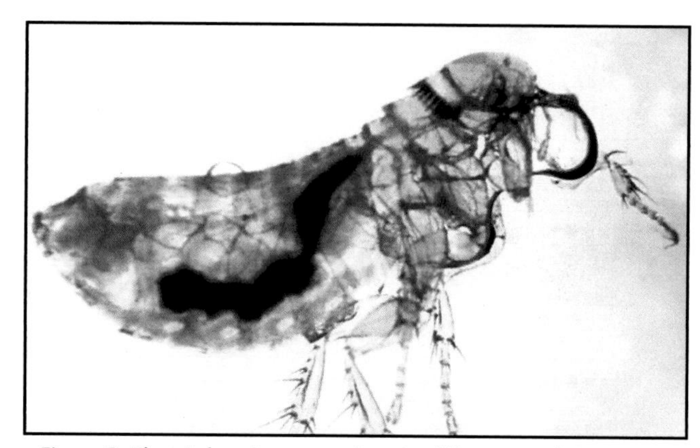

Figure 3. Thrasis bacci, a ground squirrel flea, one of the primary rodent flea vectors of plague to humans. Source: CDC

There are several forms of plague, as described below

- **Bubonic plague** is the most common form of plague, and is thought to be the type of plague that infected Europe leading in the Black Death pandemic. Bubonic plague occurs when an infected flea bites a person or when items contaminated with Y. pestis enter the body through a break in a person's skin. Bubonic plague does not spread from person to person.

- **Pneumonic plague** occurs when *Y. pestis* infects the lungs. This type of plague can spread from person to person through the air, but it usually requires direct and close contact with the ill person or animal. Pneumonic plague may also occur if a person with bubonic or septicemic plague is untreated and the bacteria spread to the lungs. Since there were no treatments available during the Black Death pandemic, it is reasonable to assume that individuals also contracted pneumonic plague at that time.
- **Septicemic plague** occurs when plague bacteria multiply in the blood. It can be a complication of pneumonic or bubonic plague or it can occur by itself. Septicemic plague does not spread from person to person.

As with many illnesses the first symptoms are broad, flu-like symptoms: fever, headache, weakness, which can rapidly develop into pneumonia with difficulty breathing and cough. Without treatment, which was the case in the 14th century, this disease has a high mortality rate. The plague is still with us today, and in fact, the World Health Organization reports 1,000 to 3,000 cases worldwide every year, with 5 to 15 of those occurring in the United States. However, the plague can be easily treated with antibiotics and if antibiotics are started within the first 24 hours, patients generally recover. There is concern today about how plague may be used as a bio-terrorism agent and the CDC home page provides interesting information on this possibility.

Clearly, the Black Death was extremely frightening for the population. It also had a long term effect on the economy and culture of people living at that time. Populations continued to decline for many decades after the epidemic. There was also increasing distrust of foreigners as people certainly feared the plague would return. When rumors circulated that Jews caused the Black Death by poisoning the wells, there were many pogroms in which Jews were killed or banished from the communities. Lepers were also persecuted. It was a painful and frightening time in the history of the world.

INFLUENZA 1918

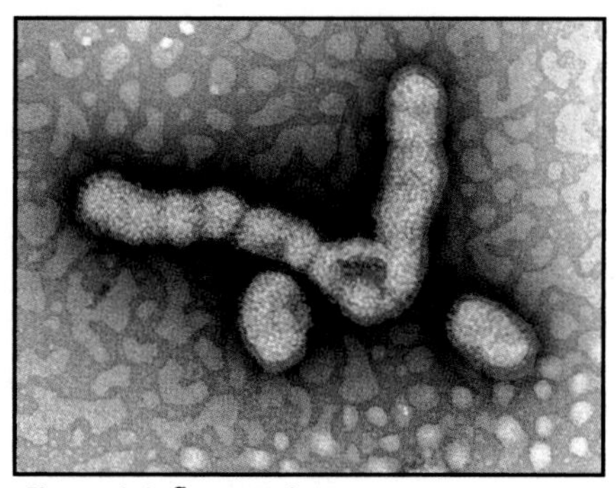

Figure 4. Influenza virus.
Source: National Institute of Allergy and Infectious Diseases (NIAID)

The United States was impacted by a major pandemic, "The Influenza Epidemic of 1918". Like the Black Death, this epidemic spread worldwide, even reaching the Arctic and remote Pacific islands, and was thus considered a pandemic. Approximately 50 million people died from this pandemic, more than had died in any other pandemic in the history of the world. It was estimated that nearly 1/3 of the world's population were infected. An unusual aspect of this influenza was that many victims were healthy young adults. Usually the elderly, less healthy individuals and infants die from influenza. The 1918 Influenza occurred at the time of World War I and, in fact was the cause of more deaths than the war. While the war did not cause the flu epidemic, it is likely that the massive troop transport and close quarters of soldiers contributed to its rapid transmission. The case fatality of the flu was >2.5% (2.5 out of every 100 people infected died) which is high compared to the usual case fatality of <0.1% (less than 1 out of every 100 people infected with flu die). It is estimated that as many as 8-10% of young adults in the world died. It is believed that this epidemic was a type of swine flu, now identified as an H1N1 influenza virus. Scientists have recovered some viral fragments from victims of the 1918 influenza and are trying to learn why it was so fatal. It is thought that perhaps elderly people had been exposed to a similar type of flu in the past and therefore had some immunity.

The 1918 influenza caused great social disruption in the country. There were no anti-viral treatments. In fact, at that time viruses had not even been seen by scientists although in 1901, Walter Reed had discovered that yellow fever was caused by a virus. And the vaccine against smallpox was developed by Edward Jenner in 1798. This epidemic occurred in the era before antibiotics (penicillin was identified in 1928) although antibiotics are not effective against viruses. Thus, many people were falling ill and there was no treatment available to cure them. Even if people didn't die, many were seriously ill and incapacitated. There were reports that many health care workers were too ill themselves to care for sick people, and likewise there were not

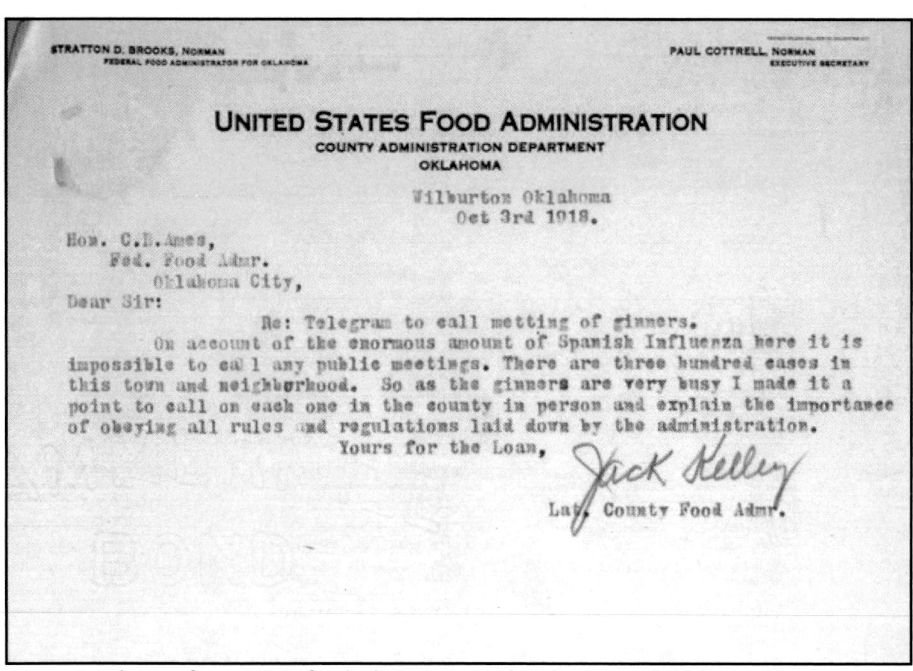

STRATTON D. BROOKS, NORMAN
FEDERAL FOOD ADMINISTRATOR FOR OKLAHOMA

PAUL COTTRELL, NORMAN
EXECUTIVE SECRETARY

UNITED STATES FOOD ADMINISTRATION
COUNTY ADMINISTRATION DEPARTMENT
OKLAHOMA

Wilburton Oklahoma
Oct 3rd 1918.

Hon. C.B. Ames,
Fed. Food Admr.
Oklahoma City,

Dear Sir:

Re: Telegram to call metting of ginners.

On account of the enormous amount of Spanish Influenza here it is impossible to call any public meetings. There are three hundred cases in this town and neighborhood. So as the ginners are very busy I made it a point to call on each one in the county in person and explain the importance of obeying all rules and regulations laid down by the administration.

Yours for the Loan,

Jack Kelley

Lat. County Food Admr.

Figure 5: Telegram from county food adminstrator to headquarters, Oaklahoma City, regarding cancellation of public meetings, October 3,1918, U.S. Food Administration.

enough gravediggers to bury the dead. Schools were closed, public gatherings were cancelled, and some communities closed all stores. Some municipalities made it a crime to shake hands. People wore masks to protect themselves, although it was not clear this was really effective. Again, try to imagine the fear of people as a severe illness affected many members of their family, killing close friends and relatives, and there was little that could be done. Some of the sick people and their children starved to death as there was no one to care for them. Healthy neighbors were too afraid of the illness to intervene. In October 1918, Victor Vaughan, Surgeon General of the Army said, "If the epidemic continues its mathematical rate of acceleration, civilization could easily disappear from the face of the earth within a few weeks."

Figure 6: Policeman in Seattle wearing masks made by the Red Cross, during the influenza epidemic, 12, 1918

Stell Altman who was 9 years old at the time and living in Rochester, New York described her experience. She, her three younger siblings, and her mother were all ill, but, miraculously, her father remained healthy and cared for his family. “I don’t know what would have happened if my father had gotten sick,” says Altman, who is now 95. “There was no help to be found anywhere; everyone was too busy caring for their own families.” Altman’s mother, Sarah Salitan, died. Her father, Morris, laid his wife’s body on a bed of straw in the home, in accord with Jewish customs. “We children didn’t go to the cemetery,” Altman says. “We were all still sick in bed.”

Could we experience a major epidemic again, and what can we do about it?

It is estimated that even with modern antiviral and antibacterial drugs, influenza vaccines, and the knowledge we have of flu prevention, we could experience another influenza virus pandemic. If this virus had a case fatality similar to the 1918 influenza, it could kill more than 100 million people worldwide. There are also other emerging diseases that could potentially mutate and cause a worldwide pandemic. There are increasing problems with antibiotic resistant pathogens, as we have had a history of overusing antibiotics. There has been an increase in the number of cases of tuberculosis, once thought to be a disease well controlled. And the AIDS epidemic has demonstrated that new epidemics can take hold and cause great mortality worldwide. But leading experts are mostly concerned about our old enemy, influenza.

Dr. Michael Osterholm, Director of the Center for Infectious Disease Research and Policy at the University of Minnesota and a leading expert in influenza preparedness, indicates that the problems resulting from a future influenza pandemic go beyond simply managing those who are ill. He states,

"The fallout from a flu pandemic could include massive energy shortages around the world, a surge in other deadly infectious diseases, uncounted associated deaths due to shortage of medical supplies and treatment, and more." He attributes much of the danger to our "just-in-time" economy in which we have relatively few stockpiles of goods and the interconnectedness of society today.

EBOLA: A WORLDWIDE PANIC

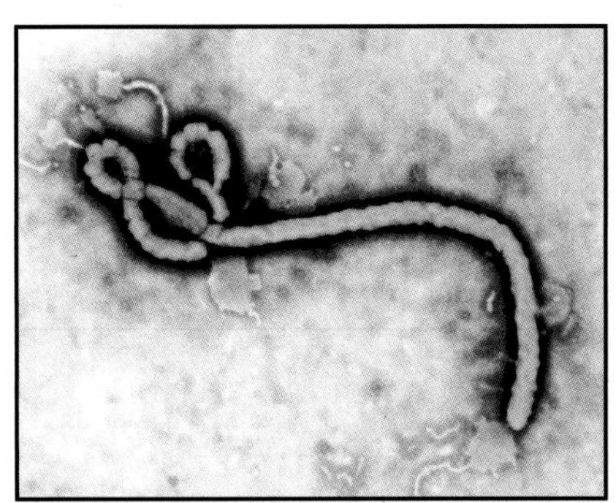

Figure 7: Ebola virus Source: CDC/ Frederick A. Murphy

In 2014, the largest epidemic of Ebola in history occurred in multiple countries in Africa, leading to worldwide panic. By the end of 2014, over 27,000 cases had been reported. While the majority of the cases remained in Africa, a few, including four in the United States (two imported and two healthcare workers), were found outside of the continent. These cases led to public health precautions not seen before including travel bans, airport screenings, and mandatory quarantine orders. Was the public overreacting or did these precautions in fact prevent an Ebola outbreak here at home?

Sign and Symptoms of Ebola

A significant part of the reason for the widespread panic can be attributed to the nature of the Ebola disease itself. Ebola is a hemorrhagic fever which by definition can lead to unexplained and uncontrollable bleeding and bruising. Other common symptoms are fever, severe headache, muscle pain, weakness, fatigue, diarrhea, vomiting, and abdominal pain. A common image of an Ebola patient is of one bleeding from multiple orifices and writhing in uncontrolled pain, which in no small part contributed to the panic of a worldwide pandemic of Ebola in 2014.

Transmission of Ebola

The original source of the Ebola disease is not confirmed, but it is widely hypothesized to have been transmitted to the first human via a diseased animal, in high probability a fruit bat or primate. While the originality of the disease is not yet understood, the method of transmission from human-to-human is well-known.

Ebola is spread through direct contact with an infected human, not airborne transmission despite the scare tactic media reporting occurring at the height of the epidemic. This direct contact infection method includes contact with blood or bodily fluids, both on the infected individual and also via touching an inanimate object in which the blood or bodily fluids of an infected individual has settled. This leads to health-care workers being at a higher risk for infection due to the fact that they are more likely to come into contact with bodily fluids while caring for an infected individual, as was the case with the two Dallas, Texas nurses who were caring for the infected Ebola patient and became infected themselves.

Another subgroup that was particularly susceptible during the 2014 epidemic was those preparing an Ebola victim's body for burial. Some areas of Africa have cultural beliefs and traditions that dictate a specific set of rituals and routines when preparing a body after the person has passed away. In some of these areas, these traditions were in direct conflict with the recommended precautions to prevent further spreading of the disease. In Sierra Leone, one of the hotspots of the epidemic, the body is traditionally handled by family and community members with bare hands as the body is washed, dressed, and buried. This cultural tradition then lead to further transmissions as "Anyone who touches a droplet of sweat, blood, or saliva from someone about to die or just deceased is at high risk of contracting the disease."(http://news.nationalgeographic.com/2015/01/150130-ebola-virus-outbreak-epidemic-sierra-leone-funerals/) Yet to ask people to abandon their beliefs and see their loved one taken away in a body bag to be cremated was an inconceivable idea and unsuccessful overall at first. People began to hide their loved ones when they were sick and dying in order to be able to safely usher their loved one through the burial ritual, inadvertently leading to further infection and death. Eventually through media campaigning, involving community leaders, and a widespread education campaign, public health workers were able to address these concerns and develop solutions that respected the cultural beliefs, but also protected the family and community members.

Controlling an Epidemic

As 2014 came to an end, the Ebola epidemic also saw significant decreases in new infections. Some tried-and-true public health measures were credited with the slowing of the infection rates. The primary method used was to "trace, isolate, and treat" all those infected. "Ebola can be stopped by tracing all the people who could have caught the disease, isolating them so they can't pass it on to others, and treating them quickly if they do develop symptoms." (http://news.nationalgeographic.com/news/2014/10/141024-ebola-nigeria-outbreak-lessons-virus-healthy/) This was initially a struggle as some communities in Africa did not believe Ebola was real and thus would not report infections and/or deaths. As various education campaigns began finding more success, the tracing and treating of those infected had a better impact on disease containment. Other containment methods utilized were to involve the local leadership, both the formal and informal community leaders and to have the public be a part of the solution process. Only through building the community's trust were the public health workers able to successfully implement the "trace, isolate, and treat" strategy to bring the Ebola Epidemic of 2014 under control.

Sources

http://www.cdc.gov/vhf/ebola/about.html

- http://www.istrianet.org/istria/medicine/infectious/1347-50_black-death1.htm
- Source: CDC Division of vector borne diseases.
- http://emergency.cdc.gov/agent/plague/factsheet.asp
- http://en.wikipedia.org/wiki/Black_death
- McCrary F and Stolley P. Examining the Plague: An Investigation of Epidemic Past and Present. Young Epidemiology Scholars Program , College Board 2004.
- http://housatonic.net/Documents/627.htm
- http://en.wikipedia.org/wiki/Flagellant
- http://www.themiddleages.net/life/decameron.html
- http://historymedren.about.com/library/weekly/aapmaps4.htm
- http://www.archives.gov/exhibits/influenza-epidemic/
- http://www.odec.ca/projects/2004/lija4j0/public_html/history.htm
- http://virus.stanford.edu/uda/fluresponse.html
- http://www.jhsph.edu/publichealthnews/articles/2005/great_influenza.html
- http://www.jhsph.edu/publichealthnews/magazine/archive/Mag_Fall04/prologues/page2.html http://forum.eastwestcenter.org/mediaconference/news-story-on-michael-osterholm-and-the-pandemic-threat/http://www.who.int./csr/disease/ebola/en/
- http://clinicalcenter.nih.gov/ebola/html
- http://news.nationalgeographic.com/2015/01/150130-ebola-virus-outbreak-epidemic-sierra-leone-funerals/
- http://news.nationalgeographic.com/news/2014/10/141024-ebola-nigeria-outbreak-lessons-virus-healthy/

Suggested Videos:

- **In Search of History - Scourge of the Black Death (History Channel) (2005)**
- **American Experience - Influenza 1918 (PBS Home Video) (2005)**
- **The New Explorers - On the Trail of a Killer Virus (A&E)(2006)**
- **Dr. Michael Osterholm Lecture: "Pandemic Influenza: Public Health's Waterloo"**
- **Frontline, Ebola Outbreak**

Link to Osterholm Lecture:

http://hscvideo2.hsc.usf.edu/asxroot/coph/EducationalMedia/deans_lecture/COPH_DLS_Dr_Osterholm_3_12_09.wvx

https://www.bopdesign.com/bop-blog/2012/10/the-elements-of-effective-brochure-design/

Assignment One: Group Brochure Assignment

You have read the synopsis on the Ebola Epidemic of 2014 in this workbook and likely followed this fascinating public health story in the media as it played out. (If you have not, I would recommend googling the Ebola Epidemic news coverage to educate yourself on how the epidemic was presented in the media.) For this assignment, you will create a brochure for the specific audience of the people of the United States.

During the non-stop Ebola media coverage in 2014, many people began to panic and feel in imminent danger. There was much inaccurate information shared nationally and overreactions by various politicians leading to further fear. Create a brochure accurately explaining the Ebola epidemic in order to alleviate some of the unnecessary concerns being felt by the public at the height of 2014. Be sure to include a definition of Ebola, methods of transmission, risks and precautions that should be taken, and any other relevant information that could be shared to calm the general public.

Creating an Effective Brochure

An effective brochure must have three important features:

1. **Be visually appealing** – The brochure must draw your audience in and make them want to look further. Choose your headlines, images, and colors carefully as they will often determine whether one reads any further
2. **Plan for the appropriate audience** – Use applicable language, information, and content for the audience you are writing to. You wouldn't use the same language to describe a disease to a group of elementary school parents as you would to a group of CDC scientists. Make sure all aspects of the brochure are appropriate to your audience.
3. **Give a powerful take-away** – The purpose of a brochure is to educate, inform, and raise a call to action. Make sure that your brochure will lead to your audience understanding your message and having the ability to take any necessary follow-up action.

For additional information on creating effective brochures, browse these sites.

http://ianrpubs.unl.edu/live/g2028/build/g2028.pdf
http://www.businessknowhow.com/directmail/ideas/brochures.htm

Module 2: Ethics in Epidemiology

Research ethics is an area of great concern. There are many challenges that an epidemiologist faces when designing and conducting a research study. There are a number of specific areas that have been identified as areas for concern: assuring autonomy of subjects, identifying risks and benefits, writing clear informed consents, and protecting the rights of all subjects, with special attention to those most vulnerable.

After reviewing the slides on ethics, see if you can identify which of the ethical principles relates to each of the following settings.

1. A researcher wants to conduct a study evaluating the association between handgun ownership and domestic violence. Because people may not be willing to participate in this type of study, he requests that the justice department provides him with the names and addresses of current gun owners so he can match them to court records.
    A. Respect for Autonomy
    B. Beneficence
    C. Justice

2. A researcher wants to conduct a study evaluating if a new medication can effectively decrease influenza symptoms. The medication needs to be administered every 8 hours for 3 days to be fully effective. It is also necessary to follow participants closely each day for one week. Early tests have not identified any side effects from the treatment. This researcher proposes to use prisoners in the study as they are exposed to influenza in a crowded system. Furthermore they can be easily followed identifying symptoms. Informed consent will be obtained from each prisoner and prisoners who agree to participate in the study will be housed in a dorm situation and moved out of individual cells. In addition, if they wish, their participation can be used as evidence of cooperative behavior in their parole hearings, potentially helping them obtain an earlier release.
    A. Respect for Autonomy
    B. Beneficence
    C. Justice

3. A researcher wants to conduct a study evaluating if a new AIDS drug can effectively prevent progression of the disease. Due to high costs of conducting the trial in the United States, she

has chosen to conduct it in a rural Ugandan community where a relatively high proportion of the people are HIV positive and it is less expensive to hire staff to monitor the participants. Participants in the study will be randomly assigned to receive this new drug or will be given a placebo. Participants will be told to continue with their routine care and they will not need to change any medications they are currently taking. It is estimated, however, that the majority of participants do not currently use any retroviral medications as they are very expensive so this will be their only opportunity to receive these drugs. The drug, which costs approximately $100 per dose, will be given to participants at no charge for 6 months. If it is successful, it can then be marketed worldwide.

A. Respect for Autonomy
B. Beneficence
C. Justice

4. A researcher wants to conduct a study to determine if a nutritional supplement improves the nutrition of children in Guatemala, as defined by an increase in weight. It is not currently known if this supplement will be effective. So the researcher randomly assigns individuals to receive this new enhanced supplement or continue taking an already established supplement that has been shown to have a small impact on weight gain. At the end of the trial if the supplement is found effective, it will be incorporated into nutritional programs for children in Guatemala.
    A. Respect for Autonomy
    B. Beneficence
    C. Justice

5. A researcher wants to conduct a study testing a new cancer treatment for individuals with lung cancer. The investigator will cover the cost of the medications, but this investigation will require additional hospitalizations which will need to be covered by the participants' insurance companies.
    A. Respect for Autonomy
    B. Beneficence
    C. Justice

Module 3: Epidemiologic measures

Epidemiologists use a number of terms to characterize exposures and disease. Some of these you will be familiar with from prior classes and some may be new. It is important to practice using the different measures so that you are familiar with them and can differentiate between them when you need to use them. There are two exercises in this chapter that should help you understand how to use these epidemiological measures.

Also, one of the main challenges of conducting epidemiological research is to determine if differences between two groups are due to chance or actually result from true differences. In the exercise in this chapter, you will learn how to do a simple analysis of data to determine this fact. In reality epidemiologists do these analyses on computers. If you are interested in learning to use a computer program to perform analysis of data, there is an excellent program available through The Centers for Disease Control and Prevention. This program is free and can be downloaded from this site:

http://www.cdc.gov/epiinfo/

Now move onto the next assignment. Part of it is to be done in your workbook and part of it is to be completed as a quiz to be provided by your instructor. Please refer to the syllabus. If you wish to practice on the computer, you can use Epi Info to calculate the Chi-square value that you will obtain in this assignment. I would suggest you also do it by hand.

You are going to be conducting your own experiment to see how epidemiologists work with rates and make decisions about differences between groups. Before you do that, you should review the information provided here on understanding the chi-square statistic. For those of you who are math phobic, just take time to look at the formulas and what they mean. Learning mathematical formulas is like learning a different language. Just as "hola" means hello in Spanish, "Σ " means Sum in statistics.

Understanding the chi-square (x^2) statistic.

Types of data: Epidemiologists use data to understand associations between risk factors and outcomes. There are several different types of data. Some data are numerical and some are categorical. Numerical data are data that can be described in a numerical form, such as your age or height. Numerical data can be continuous or it can be placed into separate categories. Categorical data are data that are placed into groups, such as gender, yes or no, and age groups. Categorical data can include items that are not numerical, e.g., race or are groups of numerical data, e.g., weight category.

Performing statistical test: When epidemiologists perform statistical tests to determine differences between variables, the type of test they use depends on the type of data. One common statistical test used is the chi-square test, which compares the counts of different responses between two or more categories. The purpose of this test is to see if the difference between the results we observed is statistically significant from what we would have expected if our hypothesis was true. In other words, this test compares our actual results with the results we would expect under our hypothesis.

Doing a chi-square test: The chi-square test is usually done on a computer but it is actually relatively easy to calculate by hand. You will calculate this test in the M&M assignment. There are three steps to this calculation:

1. You calculate the chi-square statistic.
2. You figure out the degrees of freedom.
3. You compare your results to a table to see if it is statistically significant.

Calculating the Chi-square statistic: You create a table which shows the number of items you have in each of two or more categories (called the observed number), the number of items you expect to have in each category and the total number of items. You get the expected number from some pre-existing information, for example, you might expect equal numbers of males and females so you multiply the total by 50% for each category. Or in the case of the M&M assignment the expected percent is provided by MARS company. Then you perform the following steps:

1. You subtract the expected number from the observed number for each category.
2. You square the differences to eliminate negative numbers.
3. You divide the squares for each cell by the expected number for that cell.
4. You add all these values together.

Don't worry if this seems confusing. It will be much clearer to you when you perform the chi-square test in the M&M assignment.

Degrees of freedom: The chi-square number by itself does not tell us if a result is significant as it depends on one other piece of information, the degrees of freedom. The degrees of freedom is always the number of categories minus 1. For example, if you are comparing men and women, that is two categories and so there is 1 degree of freedom. If you have three weight groups, you have 2 degrees of freedom.

The role of chance: Before any researcher conducts a study, he or she has to consider the role of chance. Researchers use samples to measure things as they usually can't measure every possible individual in the world. For example, if researchers wanted to know if a certain new medication, MR56 improves weight loss, they would compare two groups of people, those who were given MR56 and those who were given a fake pill. Researchers are not just interested in whether or not the pill works for people in their study but care about the effect of the pill on weight loss for people in general. That way if it works, drug companies can sell the pill and make lots of money. But studies can be affected by chance. It could be by chance that the group who took MR56 lived further from a pizza store and thus ate less than those who did not take the pill. So just by chance the researchers found a difference between the groups even though in reality it actually didn't work.

So, before doing a study, researchers need to decide how much chance is ok with them. Usually they set it at 5% of the time, allowing for the possibility that 5% of the time they will find a difference between the two groups when in reality there is no difference. We know groups are not perfectly equal and we also know that the larger the groups, the more likely that they will be similar. This is like flipping a coin. If you do it four times, you are more likely to have differences between the percent heads and tails but if you do it 1,000 times, it evens out and you don't have these differences.

Checking for significance: Once a researcher has a chi-square statistic and the number of degrees of freedom, he or she can just look at a table to see if their result is statistically significant. This table is in the M&M assignment. You find the row with the correct number of degrees of freedom and find the p-value for 0.05. One thing you need to notice is that as your chi-square statistic gets bigger, the probability of chance decreases. So if your number is bigger than the chi-square statistic for 0.05, then your result is statistically significant.

Once you get a chi-square statistic and you know the number of degrees of freedom, you go to a table, find the role that matches your degrees of freedom, and look at the probability for your statistic. This probability is called the p-value. If the p-value is less than 5%, then you know that it is less than 5% likely that you found a difference from chance alone and you say your answer is statistically significant.

One last thought: The 5% allowed to chance is not a hard and fast rule, but you will see it in many studies. Look at tables in research papers and they will often say <0.05%. However, if researchers do a study with many people in it and they are very worried about a false positive result, they may only allow for 1% probability of chance. Likewise, if you can only study a small number of people and a false positive is not much of a problem, you might allow 10% possibility of chance. One statistician once said not to worship at the temple of 5% chance but to think about what makes sense for your study.

M&M Assignment

This assignment will help you understand the differences between rates and proportions and how epidemiologists determine if there are differences between groups. Plus it will taste good as well. You need to complete the first part of this assignment in your workbook but save your answers. The second part is to be completed in Blackboard. Students with incorrect answers in the Blackboard part may be asked for information in the workbook. **You should complete this assignment independently.** In addition to being better able to understand how to obtain the answers, you will get to have more M&Ms for yourself.

First buy a bag of plain M&Ms. Students wonder how big a bag to get. You should know in research studies the higher the number of subjects the greater power the researcher has. Think of it like tossing a coin. Assume the chances of getting heads and tails are equal. The more times you toss it, the more likely you will get heads half the time and tails half the time. But if you only toss it twice, you will have a greater chance of getting all heads or all tails than if you toss it 100 times. So the answer in terms of power is THE BIGGER THE BETTER. It will be harder to do the assignment with a small bag from a vending machine. In terms of your overall health and waistline, you may have a different answer. You do need to get plain M&Ms, and not any special ones like Halloween or Easter M&Ms.

1. Open the bag of M&Ms but do not eat any yet. Ok you can have just 3 for now. We just talked about making graphical presentations of data. The first step is to count the subjects in your study. In this case, the subjects are the M&Ms. Separate the M&Ms into different colors and fill in the numbers and percentages in the chart below. You only need to do the first 2 rows for now. Use two decimal places for all percentages and the expected values in this assignment. It will help things add up to 100%.

Table 1. Number and percent of M&Ms in your bag as compared to the percent MARS said are in a bag.

	Brown	Yellow	Red	Green	Blue	Orange	Total
Number in your bag							
Percent in your bag							100%
% According to MARS	13%	14%	13%	16%	24%	20%	100%
Observed (o)							
Expected (e)							
Difference (o-e)							
Difference squared $(o-e)^2$							
$(o-e)^2/e$							
$X^2 = \Sigma (o-e)^2/e$							

2. Now I want you to make graphs of the colors you have. You may use POWERPOINT or EXCEL to make the graphs. You can do them by hand, but learning to use these programs is a valuable skill you will need to know in the future. So why not take the opportunity to work with your computer now. Be sure you label the graph so that a reader can understand what is in it by just looking at the picture. Look at the graphs in your book to see how they are labeled.

A. First make a bar graph and paste it in here.

B. Now make a pie chart and paste it in here. Again be sure that you label the sections, and describe what is in the figure.

We are now going to use the M&Ms to create ratios, proportions, and percentages.

Ratio

Remember, a ratio is a number created by dividing one quantity by another. A ratio compares two numbers, but the numerator does not need to be included in the denominator.

C. Write down the ratio of blue to red M&Ms.

Proportion

A proportion is a type of ratio in which the numerator is part of the denominator.

D. Write the proportion of each color using two decimal points.

Percentage

Proportions can also be shown as percentages which is a proportion x 100.

E. Show the percentage of yellow M&Ms. Be sure to include the % sign.

Rate

A rate differs from a proportion because the denominator involves a measure of time. To demonstrate this, do the following.

F. Calculate the rate of brown M&Ms that you obtained at the start of this exercise. Include the time that you identified these M&Ms, e.g., on March 1st at 5pm.

G. Now mix the M&Ms together again. Without looking at the colors, remove about 1/4 - 1/3 of the M&Ms. I will leave it up to as to how you chose to dispose of them. Eating might be a good option. Now calculate the rate of the brown M&Ms that remain at this point.

Statistical Chi Square Analysis

Before doing any statistical analysis, we need to determine the purpose of our study. Epidemiologists often write their hypotheses in the form of a null hypothesis, in which a researcher states that he or she does not expect to find a difference between two groups.

H. Write a null hypothesis below about what you expect to find.

Epidemiologists often attempt to determine if there are differences between groups. They can use the chi square statistic to determine if the differences between two groups are most likely real or just due to chance. The chi square value looks like this:

$X^2 = \Sigma (o-e)^2/e$, where Σ means sum and o = observed and e = expected.

It is less complicated than it looks if you take it step by step. We will go back to the first table and fill in some of the blanks. The observed is the number of M&Ms you counted in each color group. Look at the total number of M&Ms you have. If you had the distribution that MARS stated they had, how many would you have of each number? You can obtain this by multiplying the % by the total number of M&Ms you have and write it on the expected row. Next subtract the numbers from each other. Sometimes you will get a positive number and sometimes you will get a negative number. Because of this, the next step is to square each number and that eliminates any negative numbers. Then add these numbers up to get a total. This is your chi square value.

Write your chi square here.________

Now you have the Chi square number but you need to know what it means. You only need to compare it to a table which will help you interpret it. But before you do that, you need to determine the number of degrees of freedom in your test. The number of degrees of freedom is based upon the total number of categories you have minus 1.

Write in the number of degrees of freedom here. _________

Look at table 2 and see where your number fits in. Epidemiologists know that by chance there will be some differences between groups and they believe if the likelihood of the result occurring by chance is 5% or less, that the difference between two groups is significant. Based upon the chart, you can determine the likelihood that the amount of difference you found between your M&Ms and what MARS says is due to chance. The main thing you need to determine is if your chi square value is above or below the values for the row with the correct degrees of freedom. The main column you are concerned with is the one for 0.05. If your number is greater than the number listed in that column then most likely your result is not due to chance and you should reject the null hypothesis. If your number is less than that number then you need to fail to reject the null hypothesis. We do not talk of accepting a null hypothesis in epidemiology but rather of failing to reject it. This allows us to consider that many studies are needed before we can be confident that an exposure causes a disease. Please see the syllabus for directions on how to upload your answers on Canvas.

Table 2.

Degrees of Freedom	Probability					
	Accept Null Hypothesis				Reject Null Hypothesis	
	0.30	0.20	0.15	0.10	0.05	0.01
3	3.67	4.64	5.32	6.25	7.82	11.35
4	4.88	5.99	6.75	7.78	9.49	13.28
5	6.06	7.29	8.12	9.24	11.07	15.09
6	7.23	8.56	9.45	10.65	12.592	16.81

M&M Response Sheet

Please fill in the following. Be sure to put your answers into Canvas once you are done.

1. Number of M&Ms in your bag. ______________

2. Fill in the information from Table 1.

Number and percent of M&Ms in your bag as compared to the percent MARS said are in a bag.

	Brown	Yellow	Red	Green	Blue	Orange	Total
Number in your bag							
Percent in your bag							100%
% According to MARS	13%	14%	13%	16%	24%	20%	100%
Observed (o)							
Expected (e)							
Difference (o-e)							
Difference squared $(o-e)^2$							
$(o-e)^2/e$							
$X^2 = \Sigma (o-e)^2/e$							

3. Your chi square value. Please use two decimal places. ____________________

4. Number of degrees of freedom. ________________

5. What do you do?
   A. Fail to reject the null hypothesis
   B. Reject the null hypothesis

6. What does this mean?
   A. My M&Ms differ significantly from the number described by MARS.
   B. My M&Ms do not differ significantly from the number described by MARS.

Morbidity and Concussions in the NFL

This section of the workbook examines the morbidity and mortality is based upon the Frontline report and film: League of Denial.

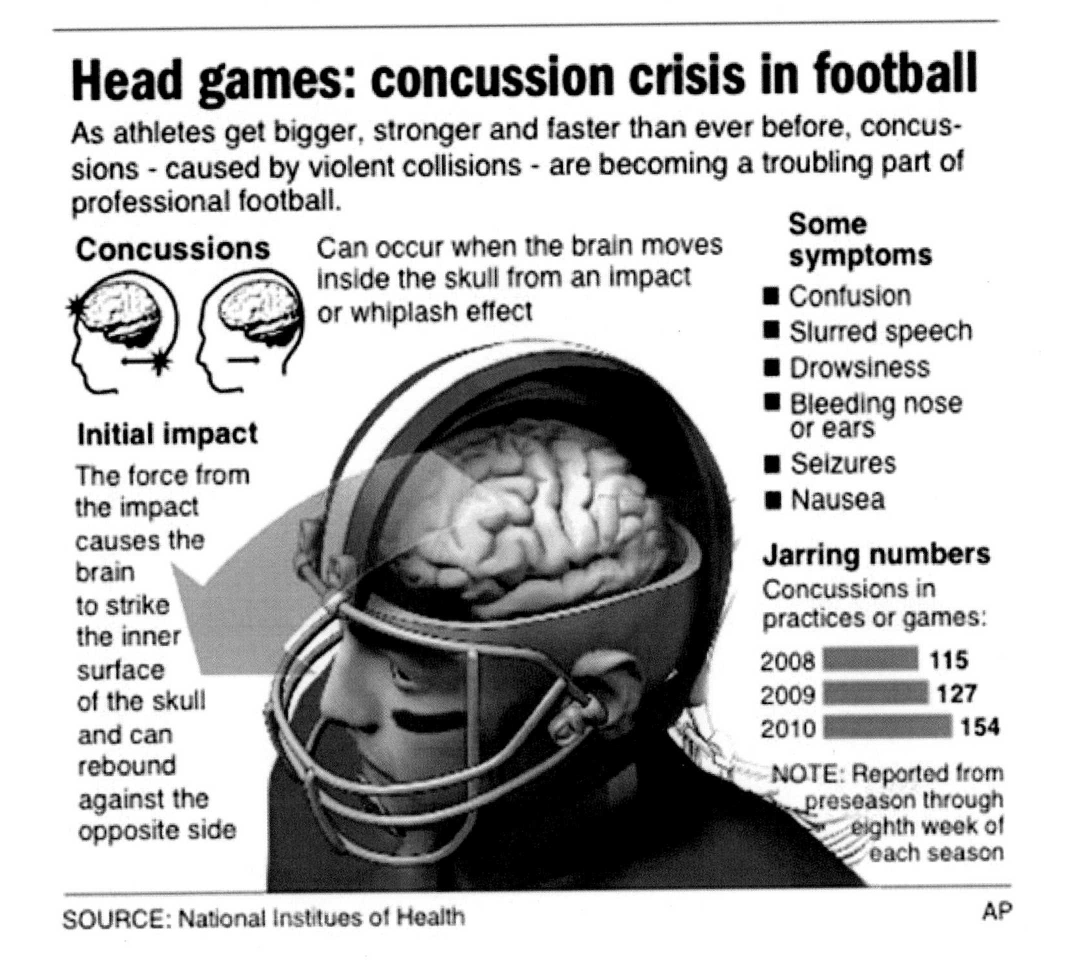

Epidemiology of Morbidity and Mortality in the NFL

Epidemiological studies evaluate the relationship of exposures and outcomes. Another term used to describe these are the independent and dependent variables. The exposure is the independent variable. Another term for exposures that you might see is risk factors. These two terms, independent variable and risk factors, are often used interchangeably. The outcome is the dependent variable as it depends on the risk factors (exposures) being studied. Before beginning a research study, it is important to identify the hypothesis or research question. In this exercise, you will be analyzing real data that describes the morbidity and mortality suffered by football players playing in the National Football League (NFL).

Traumatic Brain Injury or TBI is more commonly associated with soldiers fighting in Iraq and Afghanistan, but the term actually applies directly to injuries suffered by football players every day in the name of sport.

A concussion is a type of traumatic brain injury—or TBI—caused by a bump, blow, or jolt to the head or by a hit to the body that causes the head and brain to move rapidly back and forth. This sudden movement can cause the brain to bounce around or twist in the skull, stretching and damaging the brain cells and creating chemical changes in the brain. (www.cdc.gov)

Concussions suffered by football players result in significant morbidity and mortality, with the effects often lasting long after their career has ended. Some short-term effects linked to concussions include headaches, nausea, vomiting, dizziness, photophobia, phonophobia, balance issues, dazed and/or confused feeling, and memory problems. While many of these symptoms will self-resolve within weeks of the initial injury, "football players may be at increased risk of longer lasting cognitive deficits because of their repeated exposure to the danger". (1)

One longer-lasting effect more prevalent in football players is Post-Concussive Syndrome. PCS symptoms include the previous symptoms listed above, but also include cognitive and emotion issues including difficulty concentrating, decreased memory ability, mood disorders, including increased irritability, feeling as though one's brain is in a fog, and an overall feeling of not being "normal." These symptoms can last for days, weeks, months, or even years. In some cases, players report never fully recovering.

In 2013 at a presentation at the American Association for the Advancement of Science, long-term damage to the brains from concussive events was discussed, finding that the damage can last for decades. Dr. Maryse Lassonde, a neuropsychologist who had studies concussions in Canadian hockey players found that, "even when the symptoms of a concussion appear to have gone, the brain is still not yet 100 percent normal." Dr. Lassonde went on to discuss her findings which included the information that abnormal brain activity is evidenced for years, there is an erosion of the brain protein associated with Alzheimer's, and that neural pathways are often permanently eroded by the concussion, resulting in Parkinson-type symptoms in older players. (2)

More dangerous than a single event is a second or further concussive event, particularly if the brain has not yet healed completely from the first event. This is the epitome of the issues the NFL faced in the 2010 class-action lawsuit filed by thousands of previous NFL players. At the heart of the lawsuit was Chronic Traumatic Encephalopathy (CTE) which is "a progressive degenerative disease of the brain found in athletes (and others) with a history of repetitive brain trauma, including symptomatic concussions as well as asymptomatic subconcussive hits to the head. ...recent reports have been published of neuropathologically confirmed CTE in retired professional football players and other athletes who have a history of repetitive brain trauma." (4) As more players were diagnosed with CTE, symptoms and conditions came to light illustrating just how dangerous continuing to play with head trauma was. The lawsuit was eventually settled in 2013, with the NFL not claiming any responsibility for the injuries or events, but agreeing to monetary payments to retired players diagnosed with certain neurological conditions and a monitoring system was put into place for previous players. However, did this address the conditions for the current players? While the NFL has put new concussion protocols in place, added additional protection to the helmets, and changed some rules of legal hits, have they done enough to protect the players?

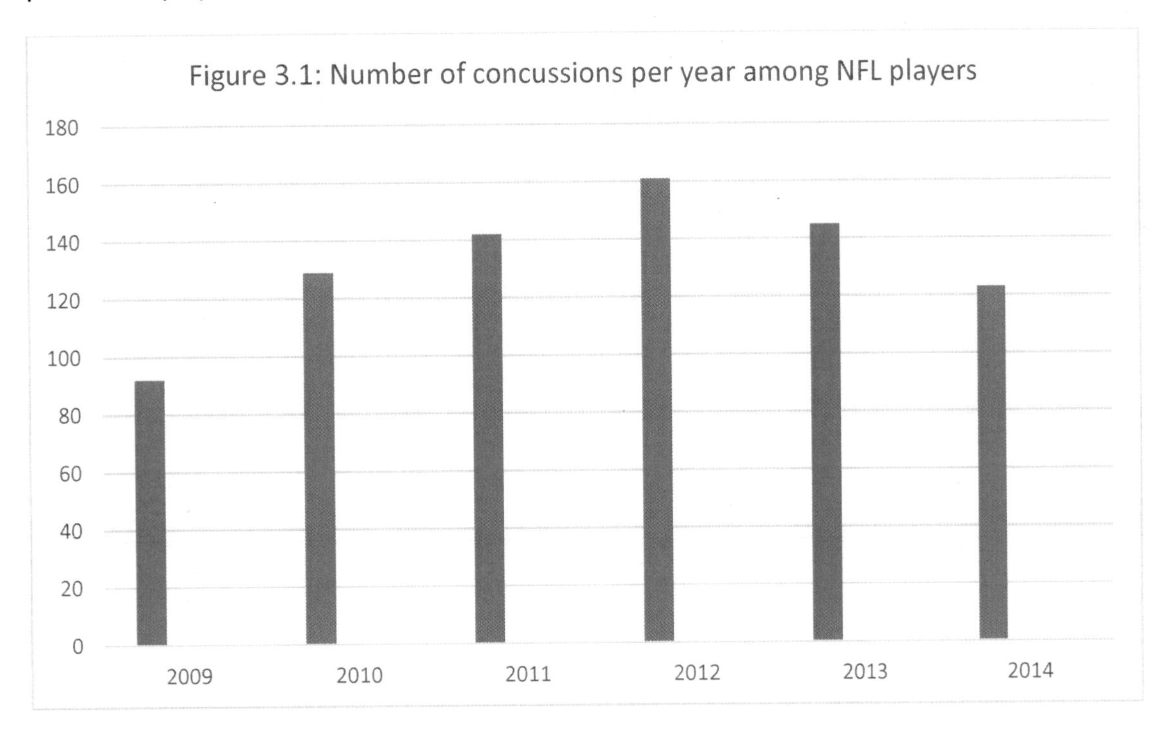

While we can see the number of concussions has decreased, there is still much concern about the health risk of playing football. Parent are worried about having their children start paying the sport, especially as their brains are potentially more at risk due to their younger ages. You will be seeing more about this topic in the future.

This section provides a number of tables with information on the numbers of concussions among football players. I want you to look at the tables to answer questions on what these tables tell us. In addition, it is sometimes much easier to understand the data if it is in a graph rather than a table. So I want you to practice making graphs.

Table 1: Offensive Concussions in the NFL 2012 - 2014

All NFL concussions	# of cases 2012	# of cases 2013	# of cases 2014	Percent of offensive concussions 2012-2014
Running Back	25	8	3	19.5%
Fullback	2	1	0	1.6%
Quarterback	9	4	0	7%
Wide Receiver	34	15	3	28.1%
Tight End	25	10	0	18.9%
Offensive Tackle	12	4	1	9.2%
Guard	19	5	0	13%
Center	2	3	0	2.7%

Table 2: Defensive Concussions in the NFL 2012 - 2014

All NFL concussions	# of cases 2012	# of cases 2013	# of cases 2014	Percent of defensive concussions 2012-2014
Cornerback	31	18	3	32.7%
Defensive End	11	7	0	11.3%
Defensive Tackle	8	2	2	7.5%
Linebacker	18	10	5	20.8%
Safety	26	13	5	27.7%

Table 3: All NFL concussions in the 2012 season

	# of concussions	# of games won in regular season	# of weeks played in playoffs
Raiders	12	4	0
Browns	9	5	0
Chiefs	9	2	0
Colts	9	11	0
Jaguars	9	2	0
Ravens	9	10	3
Cowboys	7	8	0
Jets	7	6	0
49ers	6	11	3
Packers	6	11	1
Redskins	6	10	0
Bears	5	10	0
Eagles	5	4	0
Lions	5	4	0
Patriots	5	12	2
Steelers	5	8	0
Vikings	5	10	0
Bengals	4	10	0
Broncos	4	13	1
Buccaneers	4	7	0
Cardinals	4	5	0
Giants	4	9	0
Rams	4	7	0
Saints	4	7	0
Texans	4	12	1
Titans	4	6	0
Bills	3	6	0
Panthers	3	7	0
Seahawks	3	11	1
Chargers	2	7	0
Dolphins	2	7	0
Falcolns	2	13	2

Sources

(1)http://weillcornellconcussion.org/about-concussions/long-term-effects-brain-injuries
(2)http://www.medicalnewstoday.com/articles/256518.php
(3)http://www.cdc.gov/TraumaticBrainInjury/
(4)http://www.bu.edu/cte/about/what-is-cte/
http://www.usatoday.com/story/sports/nfl/2015/04/22/nfl-concussion-lawsuit-settlement-judge-1-billion/26192827/
http://www.pbs.org/wgbh/pages/frontline/league-of-denial/ (First 3 charts)

Questions

Looking at Tables 1-3, what is (are) your outcomes(s) [dependent variable(s)? Choose all that apply.

a. Position
b. Year of play
c. Team
d. Number of concussions
e. Deaths

2. Looking at tables 1-3, what is (are) your exposure(s) [independent variable(s)? Choose all that apply.
    a. Position
    b. Year of play
    c. Team
    d. Number of concussions

3. Which offensive position had the greatest number of concussions?
    a. Running Back
    b. Fullback
    c. Quarterback
    d. Wide Receiver
    e. Tight End
    f. Offensive Tackle
    g. Guard
    h. Center

4. Which defensive position had the greatest number of concussions?
    a. Cornerback
    b. Defensive End
    c. Defensive Tackle
    d. Linebacker
    e. Safety

5. There are 16 games played by each time in a regular football season. Looking at Table 3, are teams who won at more than half of the games more likely to have concussions than those who won less than half of their games?
    a. More likely to have concussions
    b. Less likely to have concussions
    c. Have the same risk of concussions

You need to answer questions 5, by first creating your own table. Using the information from Table 3, make a list of the teams who won more than 8 games, those who won 8 games and those who won less than 8 games. Add up the number of teams in each category. Add up the number of concussions for the three groups. Fill in the table with the following information:

Table 4: Offensive Concussions in the NFL 2012 - 2014

All NFL concussions	# teams	# of concussions	# concussions per team
Won > 8 games			
Won 8 games			
Won < 8 games			
Total			

Looking at this table, you can now answer Question 5.

6. Make a graph showing the number of concussions per position for the offensive players showing the number of concussions in 2012 and 2013 for each position. Be sure to label all the parts.

7. Make a graph showing the number of concussions per position for the defensive players showing the number of concussions in 2012 and 2013 for each position. Be sure to label all the parts.

Module 4: Patterns of disease and estimation of exposure

Identifying good sources of data can be challenging. The internet is a tremendous tool in obtaining data but there is little regulation and much misinformation can be posted. One of your responsibilities as an epidemiologist is presenting accurate and using data. You can use pre-existing data or collect data yourself. Analysis of pre-existing data is called secondary analysis of data and analysis of data you collect yourself is called primary collection of data. Both are important and we will learn more about each type throughout the semester.

The type of data you use is dependent upon many factors. For many epidemiological studies, we need large datasets. The advantage is that we have enough individuals in the study to be able to determine if comparisons made are statistically significant. The disadvantage is that the data were collected for other reasons and therefore may not include measures that we are interested in.

Some good sources of reliable epidemiological data include WEB sites of trusted organizations, such as the Centers for Disease Control, the National Institute of Health, the American Public Health Association, and the World Health Organization. The U.S. census data and state and local health departments are also an excellent source of health information. Also articles published in reputable journals undergo a process called peer review. When an author submits an article to a journal, it is sent to other researchers who review the manuscript for accuracy and importance. These reviewers then inform the journal editor if the manuscript should be published. While not ideal, this system does provide a quality check. Most manuscripts in major journals can be accessed through MEDLINE or other journal search engines available at the library.

This next matching assignment will help you review some common epidemiological measures. Write in the letter of the formula that corresponds to the rate presented in the left column.

MATCHING II: Please match the following rates with the correct name by placing the appropriate formulas letter next to the rates to which they correspond.

Rates		***Formulas***
1. Crude death rates	____	(a) $\frac{\text{Number of late fetal deaths after 28 weeks Gestation + infant deaths within 7 days of birth}}{\text{Number of live births + number of late fetal deaths}}$ X 1,000
2. Case fatality rate (%)	____	(b) $\frac{\text{Number of live births within a given period}}{\text{Population size at midpoint during that period}}$ X 1,000
3. Cause specific rate	____	(c) $\frac{\text{Number of fetal deaths after 20 weeks gestation}}{\text{Number of live births + number of fetal deaths after 20 weeks or more gestation}}$ X 1,000
4. Age-specific rate	____	(d) $\frac{\text{Number of deaths from childbirth}}{\text{Number of live births}}$ X 100,000 live births (during a year)
5. Sex-specific rate	____	(e) $\frac{\text{Number of deaths in a given year}}{\text{Reference population (during midpoint of the year)}}$ X 100,000
6. Maternal mortality rate	____	(f) $\frac{\text{Mortality (or frequency of a given disease)}}{\text{Population size at midpoint of time period}}$ X 100,000
7. Fatal death rate	____	(g) $\frac{\text{Number of live births within a year}}{\text{Number of women aged 15-44 years during the midpoint of that year}}$ X 1,000 women
8. Late fetal death rate	____	(h) $\frac{\text{Number of deaths among men}}{\text{Number of men (during time period)}}$ X 100,000
9. Crude birth rate	____	(i) $\frac{\text{Number of deaths due to disease "X"}}{\text{Number of cases of disease "X"}}$ X 100 during a time period
10. General fertility rate	____	(j) $\frac{\text{Number of deaths among individuals aged 15-35 years}}{\text{Number of persons aged 15-35 years (during time period)}}$ X 100,000
11. Perinatal mortality rate	____	(k) $\frac{\text{Number of fetal deaths after 28 weeks gestation}}{\text{Number of live births + number of fetal deaths after 28 weeks or more gestation}}$ X1,000

Module 4 – Using Death Certificate Information

Death certificate information are often used to conduct research studies. When using death certificates, it is important to understand how the data are entered. This exercise gives you an opportunity to complete a death certificate so you can understand what is included and who provides the data.

Fill in the sample death certificate, using the information provided in the following description:

King Joffrey Baratheon, the eldest child of King Robert Baratheon (deceased) and Queen Cersei Lannister, died on April 13, 1467 of an apparent poisoning during his wedding reception at the Red Keep in the town of King's Landing. Witnesses at the marriage of Joffrey and Margaery Tyrell reported the king drank wine served by his uncle, Tyrion Lannister, when he began to choke and grasp for air. Joffrey fell to the ground and was declared dead at 9:47pm by the Grand Maester Pycelle who was in attendance at the wedding. According to the report filed on April 18, 1467, by coroner George R. Martin, 66 Noheart Drive, Belfast, Ireland, the autopsy showed Joffrey Baratheon died of acute cyanide poisoning.

Joffrey was born on February 16, 1448 in King's Landing in the county of the Seven Kingdoms, where he resided until his death, other than a brief period of military training provided by his father from 1442-1443 at the House of Winterfell. Joffrey listed his official occupation as king of everything. His race was listed as Caucasian. King Joffrey did not attend any years of college as he already "knew everything" upon completion of his high school years at Cruelty High School. Joffrey was legally married to Margaery Tyrell at the time of his death, with no children thankfully. Ms. Tyrell-Baratheon provided this information to complete the death certificate. The funeral director, Petyr Baelish, of Littlefinger Funeral Home in Dubrovnik was responsible for the services. King Joffrey Baratheon was buried in the Lannister Family Tombs on April 20, 1467 seven days after this death.

Insert picture

U.S. STANDARD CERTIFICATE OF DEATH

LOCAL FILE NO. STATE FILE NO.

NAME OF DECEDENT ______
For use by physician or institution

To Be Completed/ Verified By: FUNERAL DIRECTOR:

1. DECEDENT'S LEGAL NAME (Include AKA's if any) (First, Middle, Last)
2. SEX
3. SOCIAL SECURITY NUMBER

4a. AGE-Last Birthday (Years)
4b. UNDER 1 YEAR — Months / Days
4c. UNDER 1 DAY — Hours / Minutes
5. DATE OF BIRTH (Mo/Day/Yr)
6. BIRTHPLACE (City and State or Foreign Country)

7a. RESIDENCE-STATE
7b. COUNTY
7c. CITY OR TOWN
7d. STREET AND NUMBER
7e. APT. NO.
7f. ZIP CODE
7g. INSIDE CITY LIMITS? □ Yes □ No

8. EVER IN US ARMED FORCES? □ Yes □ No
9. MARITAL STATUS AT TIME OF DEATH □ Married □ Married, but separated □ Widowed □ Divorced □ Never Married □ Unknown
10. SURVIVING SPOUSE'S NAME (If wife, give name prior to first marriage)

11. FATHER'S NAME (First, Middle, Last)
12. MOTHER'S NAME PRIOR TO FIRST MARRIAGE (First, Middle, Last)

13a. INFORMANT'S NAME
13b. RELATIONSHIP TO DECEDENT
13c. MAILING ADDRESS (Street and Number, City, State, Zip Code)

14. PLACE OF DEATH (Check only one: see instructions)
IF DEATH OCCURRED IN A HOSPITAL: □ Inpatient □ Emergency Room/Outpatient □ Dead on Arrival
IF DEATH OCCURRED SOMEWHERE OTHER THAN A HOSPITAL: □ Hospice facility □ Nursing home/Long term care facility □ Decedent's home □ Other (Specify):

15. FACILITY NAME (If not institution, give street & number)
16. CITY OR TOWN , STATE, AND ZIP CODE
17. COUNTY OF DEATH

18. METHOD OF DISPOSITION: □ Burial □ Cremation □ Donation □ Entombment □ Removal from State □ Other (Specify):______
19. PLACE OF DISPOSITION (Name of cemetery, crematory, other place)

20. LOCATION-CITY, TOWN, AND STATE
21. NAME AND COMPLETE ADDRESS OF FUNERAL FACILITY

22. SIGNATURE OF FUNERAL SERVICE LICENSEE OR OTHER AGENT
23. LICENSE NUMBER (Of Licensee)

To Be Completed By: MEDICAL CERTIFIER

ITEMS 24-28 MUST BE COMPLETED BY PERSON WHO PRONOUNCES OR CERTIFIES DEATH

24. DATE PRONOUNCED DEAD (Mo/Day/Yr)
25. TIME PRONOUNCED DEAD
26. SIGNATURE OF PERSON PRONOUNCING DEATH (Only when applicable)
27. LICENSE NUMBER
28. DATE SIGNED (Mo/Day/Yr)

29. ACTUAL OR PRESUMED DATE OF DEATH (Mo/Day/Yr) (Spell Month)
30. ACTUAL OR PRESUMED TIME OF DEATH
31. WAS MEDICAL EXAMINER OR CORONER CONTACTED? □ Yes □ No

CAUSE OF DEATH (See instructions and examples)

32. **PART I.** Enter the chain of events--diseases, injuries, or complications--that directly caused the death. DO NOT enter terminal events such as cardiac arrest, respiratory arrest, or ventricular fibrillation without showing the etiology. DO NOT ABBREVIATE. Enter only one cause on a line. Add additional lines if necessary.

Approximate interval: Onset to death

IMMEDIATE CAUSE (Final disease or condition --------> resulting in death) a.______ ______
Due to (or as a consequence of):

Sequentially list conditions, if any, leading to the cause listed on line a. Enter the **UNDERLYING CAUSE** (disease or injury that initiated the events resulting in death) **LAST**
b.______ ______
Due to (or as a consequence of):
c.______ ______
Due to (or as a consequence of):
d.______ ______

PART II. Enter other significant conditions contributing to death but not resulting in the underlying cause given in PART I

33. WAS AN AUTOPSY PERFORMED? □ Yes □ No
34. WERE AUTOPSY FINDINGS AVAILABLE TO COMPLETE THE CAUSE OF DEATH? □ Yes □ No

35. DID TOBACCO USE CONTRIBUTE TO DEATH? □ Yes □ Probably □ No □ Unknown

36. IF FEMALE:
□ Not pregnant within past year
□ Pregnant at time of death
□ Not pregnant, but pregnant within 42 days of death
□ Not pregnant, but pregnant 43 days to 1 year before death
□ Unknown if pregnant within the past year

37. MANNER OF DEATH
□ Natural □ Homicide
□ Accident □ Pending Investigation
□ Suicide □ Could not be determined

38. DATE OF INJURY (Mo/Day/Yr) (Spell Month)
39. TIME OF INJURY
40. PLACE OF INJURY (e.g., Decedent's home; construction site; restaurant; wooded area)
41. INJURY AT WORK? □ Yes □ No

42. LOCATION OF INJURY: State: City or Town:
Street & Number: Apartment No.: Zip Code:

43. DESCRIBE HOW INJURY OCCURRED:

44. IF TRANSPORTATION INJURY, SPECIFY:
□ Driver/Operator
□ Passenger
□ Pedestrian
□ Other (Specify)

45. CERTIFIER (Check only one):
□ Certifying physician-To the best of my knowledge, death occurred due to the cause(s) and manner stated.
□ Pronouncing & Certifying physician-To the best of my knowledge, death occurred at the time, date, and place, and due to the cause(s) and manner stated.
□ Medical Examiner/Coroner-On the basis of examination, and/or investigation, in my opinion, death occurred at the time, date, and place, and due to the cause(s) and manner stated.

Signature of certifier:______

46. NAME, ADDRESS, AND ZIP CODE OF PERSON COMPLETING CAUSE OF DEATH (Item 32)

47. TITLE OF CERTIFIER
48. LICENSE NUMBER
49. DATE CERTIFIED (Mo/Day/Yr)
50. **FOR REGISTRAR ONLY**- DATE FILED (Mo/Day/Yr)

To Be Completed By: FUNERAL DIRECTOR

51. DECEDENT'S EDUCATION-Check the box that best describes the highest degree or level of school completed at the time of death.
□ 8th grade or less
□ 9th - 12th grade; no diploma
□ High school graduate or GED completed
□ Some college credit, but no degree
□ Associate degree (e.g., AA, AS)
□ Bachelor's degree (e.g., BA, AB, BS)
□ Master's degree (e.g., MA, MS, MEng, MEd, MSW, MBA)
□ Doctorate (e.g., PhD, EdD) or Professional degree (e.g., MD, DDS, DVM, LLB, JD)

52. DECEDENT OF HISPANIC ORIGIN? Check the box that best describes whether the decedent is Spanish/Hispanic/Latino. Check the "No" box if decedent is not Spanish/Hispanic/Latino.
□ No, not Spanish/Hispanic/Latino
□ Yes, Mexican, Mexican American, Chicano
□ Yes, Puerto Rican
□ Yes, Cuban
□ Yes, other Spanish/Hispanic/Latino (Specify) ______

53. DECEDENT'S RACE (Check one or more races to indicate what the decedent considered himself or herself to be)
□ White
□ Black or African American
□ American Indian or Alaska Native (Name of the enrolled or principal tribe) ______
□ Asian Indian
□ Chinese
□ Filipino
□ Japanese
□ Korean
□ Vietnamese
□ Other Asian (Specify)______
□ Native Hawaiian
□ Guamanian or Chamorro
□ Samoan
□ Other Pacific Islander (Specify)______
□ Other (Specify)______

54. DECEDENT'S USUAL OCCUPATION (Indicate type of work done during most of working life. DO NOT USE RETIRED).

55. KIND OF BUSINESS/INDUSTRY

MEDICAL CERTIFIER INSTRUCTIONS for selected items on U.S. Standard Certificate of Death

(See Physicians' Handbook or Medical Examiner/Coroner Handbook on Death Registration for instructions on all items)

ITEMS ON WHEN DEATH OCCURRED

Items 24-25 and 29-31 should always be completed. If the facility uses a separate pronouncer or other person to indicate that death has taken place with another person more familiar with the case completing the remainder of the medical portion of the death certificate, the pronouncer completes Items 24-28. If a certifier completes Items 24-25 as well as items 29-49, Items 26-28 may be left blank.

ITEMS 24-25, 29-30 – DATE AND TIME OF DEATH

Spell out the name of the month. If the exact date of death is unknown, enter the **approximate** date. If the date cannot be approximated, enter the date the body is found and identify as **date found**. Date pronounced and actual date may be the same. Enter the exact hour and minutes according to a 24-hour clock; estimates may be provided with "Approx." placed before the time.

ITEM 32 – CAUSE OF DEATH (See attached examples)

Take care to make the entry legible. Use a computer printer with high resolution, typewriter with good black ribbon and clean keys, or print legibly using permanent **black** ink in completing the CAUSE OF DEATH Section. **Do not abbreviate** conditions entered in section.

Part I (Chain of events leading directly to death)

•Only **one** cause should be entered on each line. Line (a) **MUST ALWAYS** have an entry. **DO NOT** leave blank. Additional lines may be added if necessary.

•If the condition on Line (a) resulted from an underlying condition, put the underlying condition on Line (b), and so on, until the full sequence is reported. **ALWAYS** enter the **underlying cause of death** on the lowest used line in Part I.

•For each cause indicate the best estimate of the interval between the presumed onset and the date of death. The terms "unknown" or "approximately" may be used. General terms, such as minutes, hours, or days, are acceptable, if necessary. **DO NOT** leave blank.

•The terminal event (for example, cardiac arrest or respiratory arrest) should not be used. If a mechanism of death seems most appropriate to you for line (a), then you must always list its cause(s) on the line(s) below it (for example, cardiac arrest **due to** coronary artery atherosclerosis *or* cardiac arrest **due to** blunt impact to chest).

• If an organ system failure such as congestive heart failure, hepatic failure, renal failure, or respiratory failure is listed as a cause of death, always report its etiology on the line(s) beneath it (for example, renal failure **due to** Type I diabetes mellitus).

•When indicating neoplasms as a cause of death, include the following: 1) primary site *or* that the primary site is unknown, 2) benign or malignant, 3) cell type *or* that the cell type is unknown, 4) grade of neoplasm, and 5) part or lobe of organ affected. (For example, a primary well-differentiated squamous cell carcinoma, lung, left upper lobe.)

•Always report the fatal injury (for example, stab wound of chest), the trauma (for example, transection of subclavian vein), and impairment of function (for example, air embolism).

PART II (Other significant conditions)

•Enter all diseases or conditions contributing to death that were not reported in the chain of events in Part I and that did not result in the **underlying cause of death**. See attached examples.

•If two or more possible sequences resulted in death, or if two conditions seem to have added together, report in Part I the one that, in your opinion, most directly caused death. Report in Part II the other conditions or diseases.

CHANGES TO CAUSE OF DEATH

Should additional medical information or autopsy findings become available that would change the cause of death originally reported, the original death certificate should be amended by the certifying physician by **immediately** reporting the revised cause of death to the State Vital Records Office.

ITEMS 33-34 - AUTOPSY

•33 - Enter "Yes" if either a partial or full autopsy was performed. Otherwise enter "No."

•34 - Enter "Yes" if autopsy findings were available to complete the cause of death; otherwise enter "No". Leave item blank if no autopsy was performed.

ITEM 35 - DID TOBACCO USE CONTRIBUTE TO DEATH?

Check "yes" if, in your opinion, the use of tobacco contributed to death. Tobacco use may contribute to deaths due to a wide variety of diseases; for example, tobacco use contributes to many deaths due to emphysema or lung cancer and some heart disease and cancers of the head and neck. Check "no" if, in your clinical judgment, tobacco use did not contribute to this particular death.

ITEM 36 - IF FEMALE, WAS DECEDENT PREGNANT AT TIME OF DEATH OR WITHIN PAST YEAR?

This information is important in determining pregnancy-related mortality.

ITEM 37 - MANNER OF DEATH

•Always check Manner of Death, which is important: 1) in determining accurate causes of death; 2) in processing insurance claims; and 3) in statistical studies of injuries and death.

•Indicate "Pending investigation" if the manner of death cannot be determined whether due to an accident, suicide, or homicide within the statutory time limit for filing the death certificate. This should be changed later to one of the other terms.

•Indicate "Could not be Determined" **ONLY** when it is impossible to determine the manner of death.

ITEMS 38-44 - ACCIDENT OR INJURY – to be filled out in all cases of deaths due to injury or poisoning.

•38 - Enter the exact month, day, and year of injury. Spell out the name of the month. **DO NOT** use a number for the month. (Remember, the date of injury may differ from the date of death.) Estimates may be provided with "Approx." placed before the date.

•39 - Enter the exact hour and minutes of injury or use your best estimate. Use a 24-hour clock.

•40 - Enter the general place (such as restaurant, vacant lot, or home) where the injury occurred. **DO NOT** enter firm or organization names. (For example, enter "factory", **not** "Standard Manufacturing, Inc.")

•41 - Complete if anything other than natural disease is mentioned in Part I or Part II of the medical certification, including homicides, suicides, and accidents. This includes all motor vehicle deaths. The item **must** be completed for decedents ages 14 years or over and may be completed for those less than 14 years of age if warranted. Enter "Yes" if the injury occurred at work. Otherwise enter "No". An injury may occur at work regardless of whether the injury occurred in the course of the decedent's "usual" occupation. Examples of injury at work and injury not at work follow:

Injury at work	**Injury not at work**
Injury while working or in vocational training on job premises	Injury while engaged in personal recreational activity on job premises
Injury while on break or at lunch or in parking lot on job premises	Injury while a visitor (not on official work business) to job premises
Injury while working for pay or compensation, including at home	Homemaker working at homemaking activities
Injury while working as a volunteer law enforcement official etc.	Student in school
Injury while traveling on business, including to/from business contacts	Working for self for no profit (mowing yard, repairing own roof, hobby)
	Commuting to or from work

•42 - Enter the complete address where the injury occurred including zip code.

•43 - Enter a brief but specific and clear description of how the injury occurred. Explain the circumstances or cause of the injury. Specify **type of gun** or **type of vehicle** (e.g., car, bulldozer, train, etc.) when relevant to circumstances. Indicate if more than one vehicle involved; specify type of vehicle decedent was in.

•44 -Specify role of decedent (e.g. driver, passenger). Driver/operator and passenger should be designated for modes other than motor vehicles such as bicycles. Other applies to watercraft, aircraft, animal, or people attached to outside of vehicles (e.g. surfers).

Rationale: Motor vehicle accidents are a major cause of unintentional deaths; details will help determine effectiveness of current safety features and laws.

REFERENCES

For more information on how to complete the medical certification section of the death certificate, refer to tutorial at http://www.TheNAME.org and resources including instructions and handbooks available by request from NCHS, Room 7318, 3311 Toledo Road, Hyattsville, Maryland 20782-2003 or at www.cdc.gov/nchs/about/major/dvs/handbk.htm

Cause-of-death – Background, Examples, and Common Problems

Accurate cause of death information is important
•to the public health community in evaluating and improving the health of all citizens, and
•often to the family, now and in the future, and to the person settling the decedent's estate.

The cause-of-death section consists of two parts. **Part I** is for reporting a chain of events leading directly to death, with the **immediate cause** of death (the final disease, injury, or complication directly causing death) on line a and the **underlying cause** of death (the disease or injury that initiated the chain of events that led directly and inevitably to death) on the lowest used line. **Part II** is for reporting all other significant diseases, conditions, or injuries that contributed to death but which did not result in the underlying cause of death given in **Part I**. **The cause-of-death information should be YOUR best medical OPINION.** A condition can be listed as "probable" even if it has not been definitively diagnosed.

Examples of properly completed medical certifications

CAUSE OF DEATH (See instructions and examples)

32. **PART I.** Enter the chain of events--diseases, injuries, or complications--that directly caused the death. DO NOT enter terminal events such as cardiac arrest, respiratory arrest, or ventricular fibrillation without showing the etiology. DO NOT ABBREVIATE. Enter only one cause on a line. Add additional lines if necessary.

		Approximate interval: Onset to death
IMMEDIATE CAUSE (Final disease or condition ---------> resulting in death)	a. Rupture of myocardium Due to (or as a consequence of):	Minutes
Sequentially list conditions, if any, leading to the cause listed on line a. Enter the **UNDERLYING CAUSE** (disease or injury that initiated the events resulting in death) **LAST**	b. Acute myocardial infarction Due to (or as a consequence of):	6 days
	c. Coronary artery thrombosis Due to (or as a consequence of):	5 years
	d. Atherosclerotic coronary artery disease	7 years

PART II. Enter other significant conditions contributing to death but not resulting in the underlying cause given in PART I

Diabetes, Chronic obstructive pulmonary disease, smoking

33. WAS AN AUTOPSY PERFORMED? ■ Yes □ No

34. WERE AUTOPSY FINDINGS AVAILABLE TO COMPLETE THE CAUSE OF DEATH? ■ Yes □ No

35. DID TOBACCO USE CONTRIBUTE TO DEATH? ■ Yes □ Probably □ No □ Unknown

36. IF FEMALE: ■ Not pregnant within past year □ Pregnant at time of death □ Not pregnant, but pregnant within 42 days of death □ Not pregnant, but pregnant 43 days to 1 year before death □ Unknown if pregnant within the past year

37. MANNER OF DEATH ■ Natural □ Homicide □ Accident □ Pending Investigation □ Suicide □ Could not be determined

CAUSE OF DEATH (See instructions and examples)

32. **PART I.** Enter the chain of events--diseases, injuries, or complications--that directly caused the death. DO NOT enter terminal events such as cardiac arrest, respiratory arrest, or ventricular fibrillation without showing the etiology. DO NOT ABBREVIATE. Enter only one cause on a line. Add additional lines if necessary.

		Approximate interval: Onset to death
IMMEDIATE CAUSE (Final disease or condition ---------> resulting in death)	a. Aspiration pneumonia Due to (or as a consequence of):	2 Days
Sequentially list conditions, if any, leading to the cause listed on line a. Enter the **UNDERLYING CAUSE** (disease or injury that initiated the events resulting in death) **LAST**	b. Complications of coma Due to (or as a consequence of):	7 weeks
	c. Blunt force injuries Due to (or as a consequence of):	7 weeks
	d. Motor vehicle accident	7 weeks

PART II. Enter other significant conditions contributing to death but not resulting in the underlying cause given in PART I

33. WAS AN AUTOPSY PERFORMED? ■ Yes □ No

34. WERE AUTOPSY FINDINGS AVAILABLE TO COMPLETE THE CAUSE OF DEATH? ■ Yes □ No

35. DID TOBACCO USE CONTRIBUTE TO DEATH? □ Yes □ Probably ■ No □ Unknown

36. IF FEMALE: □ Not pregnant within past year □ Pregnant at time of death □ Not pregnant, but pregnant within 42 days of death □ Not pregnant, but pregnant 43 days to 1 year before death □ Unknown if pregnant within the past year

37. MANNER OF DEATH □ Natural □ Homicide ■ Accident □ Pending Investigation □ Suicide □ Could not be determined

38. DATE OF INJURY (Mo/Day/Yr) (Spell Month)	39. TIME OF INJURY	40. PLACE OF INJURY (e.g., Decedent's home; construction site; restaurant; wooded area)	41. INJURY AT WORK?
August 15, 2003	Approx. 2320	road side near state highway	□ Yes ■ No

42. LOCATION OF INJURY: State: Missouri City or Town: near Alexandria
Street & Number: mile marker 17 on state route 46a Apartment No.: Zip Code:

43. DESCRIBE HOW INJURY OCCURRED:
Decedent driver of van, ran off road into tree

44. IF TRANSPORTATION INJURY, SPECIFY: ■ Driver/Operator □ Passenger □ Pedestrian □ Other (Specify)

Common problems in death certification

The **elderly decedent** should have a clear and distinct etiological sequence for cause of death, if possible. Terms such as senescence, infirmity, old age, and advanced age have little value for public health or medical research. Age is recorded elsewhere on the certificate. When a number of conditions resulted in death, the physician should choose the single sequence that, in his or her opinion, best describes the process leading to death, and place any other pertinent conditions in Part II. If after careful consideration the physician cannot determine a sequence that ends in death, then the medical examiner or coroner should be consulted about conducting an investigation or providing assistance in completing the cause of death.

The **infant decedent** should have a clear and distinct etiological sequence for cause of death, if possible. "Prematurity" should not be entered without explaining the etiology of prematurity. Maternal conditions may have initiated or affected the sequence that resulted in infant death, and such maternal causes should be reported in addition to the infant causes on the infant's death certificate (e.g., Hyaline membrane disease **due to** prematurity, 28 weeks **due to** placental abruption **due to** blunt trauma to mother's abdomen).

When **SIDS** is suspected, a complete investigation should be conducted, typically by a medical examiner or coroner. If the infant is under 1 year of age, no cause of death is determined after scene investigation, clinical history is reviewed, and a complete autopsy is performed, then the death can be reported as Sudden Infant Death Syndrome.

When processes such as the following are reported, additional information about the etiology should be reported:

Abscess
Abdominal hemorrhage
Adhesions
Adult respiratory distress syndrome
Acute myocardial infarction
Altered mental status
Anemia
Anoxia
Anoxic encephalopathy
Arrhythmia
Ascites
Aspiration
Atrial fibrillation
Bacteremia
Bedridden
Biliary obstruction
Bowel obstruction
Brain injury
Brain stem herniation
Carcinogenesis
Carcinomatosis
Cardiac arrest
Cardiac dysrhythmia
Cardiomyopathy
Cardiopulmonary arrest
Cellulitis
Cerebral edema
Cerebrovascular accident
Cerebellar tonsillar herniation
Chronic bedridden state
Cirrhosis
Coagulopathy
Compression fracture
Congestive heart failure
Convulsions
Decubiti
Dehydration
Dementia (when not otherwise specified)
Diarrhea
Disseminated intra vascular coagulopathy
Dysrhythmia
End-stage liver disease
End-stage renal disease
Epidural hematoma
Exsanguination
Failure to thrive
Fracture
Gangrene
Gastrointestinal hemorrhage
Heart failure
Hemothorax
Hepatic failure
Hepatitis
Hepatorenal syndrome
Hyperglycemia
Hyperkalemia
Hypovolemic shock
Hyponatremia
Hypotension
Immunosuppression
Increased intra cranial pressure
Intra cranial hemorrhage
Malnutrition
Metabolic encephalopathy
Multi-organ failure
Multi-system organ failure
Myocardial infarction
Necrotizing soft-tissue infection
Old age
Open (or closed) head injury
Paralysis
Pancytopenia
Perforated gallbladder
Peritonitis
Pleural effusions
Pneumonia
Pulmonary arrest
Pulmonary edema
Pulmonary embolism
Pulmonary insufficiency
Renal failure
Respiratory arrest
Seizures
Sepsis
Septic shock
Shock
Starvation
Subdural hematoma
Subarachnoid hemorrhage
Sudden death
Thrombocytopenia
Uncal herniation
Urinary tract infection
Ventricular fibrillation
Ventricular tachycardia
Volume depletion

If the certifier is unable to determine the etiology of a process such as those shown above, the process must be qualified as being of an unknown, undetermined, probable, presumed, or unspecified etiology so it is clear that a distinct etiology was not inadvertently or carelessly omitted.

The following conditions and types of death might seem to be specific or natural but when the medical history is examined further may be found to be complications of an injury or poisoning (possibly occurring long ago). Such cases should be reported to the medical examiner/coroner.

Asphyxia
Bolus
Choking
Drug or alcohol overdose/drug or alcohol abuse
Epidural hematoma
Exsanguination
Fall
Fracture
Hip fracture
Hyperthermia
Hypothermia
Open reduction of fracture
Pulmonary emboli
Seizure disorder
Sepsis
Subarachnoid hemorrhage
Subdural hematoma
Surgery
Thermal burns/chemical burns

REV. 11/2003

FUNERAL DIRECTOR INSTRUCTIONS for selected items on U.S.

Standard Certificate of Death (For additional information concerning all items on certificate see Funeral Directors' Handbook on Death Registration)

ITEM 1. DECEDENT'S LEGAL NAME
Include any other names used by decedent, if substantially different from the legal name, after the abbreviation AKA (also known as) e.g. Samuel Langhorne Clemens AKA Mark Twain, **but not** Jonathon Doe AKA John Doe

ITEM 5. DATE OF BIRTH
Enter the full name of the month (January, February, March etc.) Do not use a number or abbreviation to designate the month.

ITEM 7A-G. RESIDENCE OF DECEDENT (information divided into seven categories)
Residence of decedent is the place where the decedent actually resided. The place of residence is not necessarily the same as "home state" or "legal residence". Never enter a temporary residence such as one used during a visit, business trip, or vacation. Place of residence during a tour of military duty or during attendance at college is considered permanent and should be entered as the place of residence. If the decedent had been living in a facility where an individual usually resides for a long period of time, such as a group home, mental institution, nursing home, penitentiary, or hospital for the chronically ill, report the location of that facility in item 7. If the decedent was an infant who never resided at home, the place of residence is that of the parent(s) or legal guardian. **Never** use an acute care hospital's location as the place of residence for any infant. If Canadian residence, please specify Province instead of State.

ITEM 10. SURVIVING SPOUSE'S NAME
If the decedent was married at the time of death, enter the full name of the surviving spouse. If the surviving spouse is the wife, enter her name prior to first marriage. This item is used in establishing proper insurance settlements and other survivor benefits.

ITEM 12. MOTHER'S NAME PRIOR TO FIRST MARRIAGE
Enter the name used prior to first marriage, commonly known as the maiden name. This name is useful because it remains constant throughout life.

ITEM 14. PLACE OF DEATH
The place where death is pronounced should be considered the place where death occurred. If the place of death is unknown but the body is found in your State, the certificate of death should be completed and filed in accordance with the laws of your State. Enter the place where the body is found as the place of death.

ITEM 51. DECEDENT'S EDUCATION *(Check appropriate box on death certificate)*
Check the box that corresponds to the highest level of education that the decedent completed. **Information in this section will not appear on the certified copy of the death certificate. This information is used to study the relationship between mortality and education (which roughly corresponds with socioeconomic status). This information is valuable in medical studies of causes of death and in programs to prevent illness and death.**

ITEM 52. WAS DECEDENT OF HISPANIC ORIGIN? *(Check "No" or appropriate "Yes" box)*
Check "No" or check the "Yes" box that best corresponds with the decedent's ethnic Spanish identity as given by the informant. Note that "Hispanic" is not a race and item 53 must also be completed. Do not leave this item blank. With respect to this item, "Hispanic" refers to people whose origins are from Spain, Mexico, or the Spanish-speaking Caribbean Islands or countries of Central or South America. Origin includes ancestry, nationality, and lineage. There is no set rule about how many generations are to be taken into account in determining Hispanic origin; it may be based on the country of origin of a parent, grandparent, or some far-removed ancestor. Although the prompts include the major Hispanic groups, other groups may be specified under "other". "Other" may also be used for decedents of multiple Hispanic origin (e.g. Mexican-Puerto Rican). **Information in this section will not appear on the certified copy of the death certificate. This information is needed to identify health problems in a large minority population in the United States. Identifying health problems will make it possible to target public health resources to this important segment of our population.**

ITEM 53. RACE *(Check appropriate box or boxes on death certificate)*
Enter the race of the decedent as stated by the informant. Hispanic is not a race; information on Hispanic ethnicity is collected separately in item 52. American Indian and Alaska Native refer only to those native to North and South America (including Central America) and does not include Asian Indian. Please specify the name of enrolled or principal tribe (e.g., Navajo, Cheyenne, etc.) for the American Indian or Alaska Native. For Asians check Asian Indian, Chinese, Filipino, Japanese, Korean, Vietnamese, or specify other Asian group; for Pacific Islanders check Guamanian or Chamorro, Samoan, or specify other Pacific Island group. If the decedent was of mixed race, enter each race (e.g., Samoan-Chinese-Filipino or White, American Indian). **Information in this section will not appear on the certified copy of the death certificate. Race is essential for identifying specific mortality patterns and leading causes of death among different racial groups. It is also used to determine if specific health programs are needed in particular areas and to make population estimates.**

ITEMS 54 AND 55. OCCUPATION AND INDUSTRY
Questions concerning occupation and industry must be completed for all decedents 14 years of age or older. This information is useful in studying deaths related to jobs and in identifying any new risks. For example, the link between lung disease and lung cancer and asbestos exposure in jobs such as shipbuilding or construction was made possible by this sort of information on death certificates. **Information in this section will not appear on the certified copy of the death certificate.**

ITEM 54. DECEDENT'S USUAL OCCUPATION
Enter the usual occupation of the decedent. This is not necessarily the last occupation of the decedent. Never enter "retired". Give kind of work decedent did during most of his or her working life, such as claim adjuster, farmhand, coal miner, janitor, store manager, college professor, or civil engineer. If the decedent was a homemaker at the time of death but had worked outside the household during his or her working life, enter that occupation. If the decedent was a homemaker during most of his or her working life, and never worked outside the household, enter "homemaker". Enter "student" if the decedent was a student at the time of death and was never regularly employed or employed full time during his or her working life. **Information in this section will not appear on the certified copy of the death certificate.**

ITEM 55. KIND OF BUSINESS/INDUSTRY
Kind of business to which occupation in item 54 is related, such as insurance, farming, coal mining, hardware store, retail clothing, university, or government. DO NOT enter firm or organization names. If decedent was a homemaker as indicated in item 54, then enter either "own home" or "someone else's home" as appropriate. If decedent was a student as indicated in item 54, then enter type of school, such as high school or college, in item 55. **Information in this section will not appear on the certified copy of the death certificate.**

NOTE: This recommended standard death certificate is the result of an extensive evaluation process. Information on the process and resulting recommendations as well as plans for future activities is available on the Internet at: http://www.cdc.gov/nchs/vital_certs_rev.htm.

REV. 11/2003

Risk Factors for Pneumonia Mortality

Death certificates are used in many studies to identify trends in mortality as well as obtain information on causes of deaths. The following exercise demonstrates how death certificate data can be used. As you saw in the previous exercise, death certificates include demographic information, the date and time of death, as well as both the immediate and underlying causes of death. It is important to consider both the immediate cause of death as well as underlying causes of death when using death certificate data. This is because the immediate cause of death may be an event that resulted from an underlying condition. For example, a person with AIDS who dies of cardiac arrest may have cardiac arrest listed as the immediate cause of death and AIDS listed as an underlying cause of death. The causes of death are recorded as numerical values based upon ICD9 or ICD 10 codes.

ICD-9 Codes: According to the CDC[1], "The International Classification of Diseases (ICD) is the classification used to code and classify mortality data from death certificates. The International Classification of Diseases, Clinical Modification is used to code and classify morbidity data from the inpatient and outpatient records, physician offices, and most National Center for Health Statistics (NCHS) surveys." While these codes were updated and ICD 10 codes are now in use, this exercise uses the ICD 9 codes.

Death Certificate Exercise

During a recent flu epidemic there were increased numbers of deaths due to pneumonia, which was noted in media reports. To evaluate the situation, you decide to review the mortality data for the prior year. The data includes the following: information from the death certificates on **573 deaths**, all assigned ICD9 codes and v-codes associated with these deaths, and the demographic information that is available on the death certificates.

Table 1. ICD-9 codes for pneumonia and influenza (480-488)[2]

- (480) Viral pneumonia
    - (480.31) Pneumonia, SARS associated coronavirus
  - (480.9) Pneumonia, viral, unspec.
- (481) Pneumococcal pneumonia
- (482) Other bacterial pneumonia
  - (482.9) Pneumonia, bacterial, unspec.
- (483) Pneumonia due to other specified organism
  - (483.0) Mycoplasma pneumoniae
- (485) Bronchopneumonia, organism unspecified
- (486) Pneumonia, organism unspecified
- (487) Influenza
  - (487.0) Influenza w/ pneumonia
  - (487.1) Influenza w/ other respiratory manifestations
- (488) Influenza due to identified Avian influenza virus
- (488.1) Influenza due to identified 2009 H1N1 virus

Here is the following additional relevant information for the community under study:

78 patients with ICD 9 codes #480-486 died in 2005

2,600 patients with ICD 9 codes #480-486 were discharged alive from the hospital in 2005

43 patients with ICD 9 codes #487-488 died in 2005

47 patients with ICD 9 codes #487-488 were discharged alive from the hospital in 2005

Overall population of the community at midpoint = 165,378.

Answer the following questions using these mortality data

Which of the following information is available in the death certificate? For each category indicate Y for Yes or N for No.

Table 1. Variables included on death certificates.

Variable	Response
1. Age	Y N
2. Immediate cause of death	Y N
3. Pregnancy status	Y N
4. Birthplace	Y N
5. Religion	Y N
6. Marital status	Y N
7. Military veteran	Y N
8. Number of children	Y N
9. Educational level	Y N
10. Tobacco use contributed to death	Y N

11. What is the crude mortality rate for this community?

12. What is the proportional mortality due to pneumonia in your file?

13. While the proportional mortality rate seems fairly high, you know that the proportional mortality rate can be affected by several things. Which of the following factors can impact on a proportional mortality rate?
    A. Decrease in mortality from other causes
    B. Increase in mortality from the cause being studied
    C. A and B

14. What is the case fatality among patients hospitalized with influenza during the study period.

15. Does this mean that influenza has a high case-fatality rate during this study period?
    A. Not necessarily as this only includes patients with influenza who were hospitalized that year and it is likely that sicker patients are in the hospital
    B. Yes, this number indicates that almost 50% of individuals with influenza die
    C. Not necessarily as the number of people dying from other causes may be lower this year than before
    D. A and C

Table 2 presents information on the average age of individuals with ICD 9 codes for pneumonia and influenza.

Table 2. Average age of individuals with deaths recorded using ICD9 Codes for pneumonia, influenza and both, by gender.

	Average Age		
ICD 9 Codes	Men	Women	p-value
Pneumonia only	88	86	0.76
Influenza only	65	59	0.12
Both pneumonia and influenza	64	54	0.02

Although the age distribution is similar for men and women with either pneumonia or influenza, women in this database who have ICD9 codes for both pneumonia and influenza are significantly younger than men who had both codes. (A p-value of <0.05 means that the difference between the two groups was statistically significant. You will learn more about p-values later.)

Table 3 presents the association of current pregnancy for individuals with ICD 9 codes for pneumonia and influenza.

Table 3. Percent of pregnant women with deaths recorded using ICD9 codes for pneumonia, influenza and both.

ICD 9 Codes	% pregnant
Pneumonia only	0.4%
Influenza only	1.3%
Both pneumonia and influenza	3.2%

P=0.04

16. How would these findings be interpreted?
    A. Being pregnant causes women to die of influenza
    B. There is evidence of an association between influenza mortality and pregnancy

17. Which of the following is true of this study?
    A. Death certificate analysis can help identify associations
    B. Based on this study, a national warning should be issued informing women that if they are pregnant they are at greater risk of influenza mortality
    C. Analyzing death certificates can identify the causes of disease

18. Which of the following cannot be determined from using death certificates
    A. The time of death
    B. The trimester of pregnancy in which the death occurred
    C. Current marital status
    D. Age at which smoking started
    E. B and D

Sources:

1. http://www.cdc.gov/nchs/icd.htm
2. http://en.wikipedia.org/wiki/List_of_ICD-9_codes
2. http://en.wikipedia.org/wiki/ICD-9_V_codes
4. http://www.cdc.gov/h1n1flu/pdf/tip_sheet_pregnant.pdf

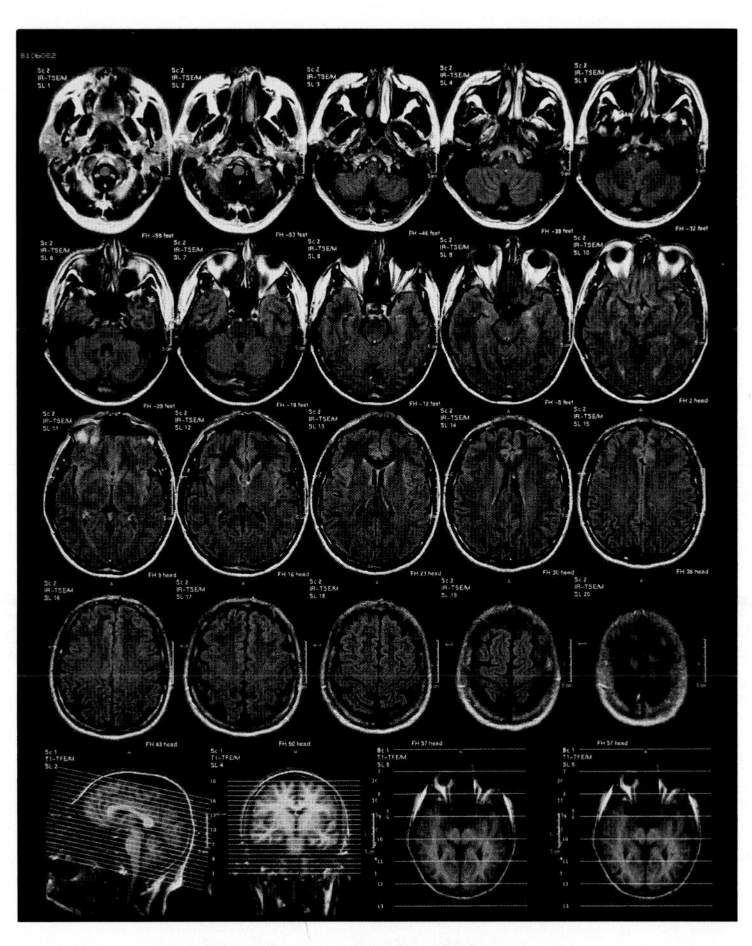

Module 5: Descriptive Epidemiology and causation

Epidemiologists often have to rely on data collected for other purposes when answering a research question. One example of this is birth certificate data. Most countries routinely collect information on vital records of their populations: births, deaths, marriages, etc. In the United States, every state collects information on the births in that state, and that information is relayed to the National Center for Health Statistics. Information from the birth certificates is used to the number and rates of births, identify risk factors for certain conditions, and can be used for a variety of program planning activities. Birth certificates are considered confidential and are not released to the general public so the following exercise uses mock certificates.

Birth Certificate Exercise

You are working as an intern for a local health agency and have been contacted by the local newspaper for background on an article on teenage pregnancy. People have been concerned that there may be high rates of teenage pregnancy in the city. The previous intern was looking at this issue and he had already reviewed most of the birth certificates for the Year 2011, but he had some he was not certain about. You have been asked to review these remaining 51 certificates and determine how many of the questionable certificates were teenage births in Tampa, FL during 2011.

Before you answer this question, you need to determine a case definition. A case definition is a set of rules that guide how you identify a case in an epidemiological investigation. If you are interested in teenage births in Tampa, FL, you need to decide on a set of rules when evaluating birth certificates.

1. The first rule is age. You should use the age of teenage births, as defined by the National Center for Health Statistics. The reason for using this age is that you can compare your results to the national data. What is the definition of teenage birth by the National Center for Health Statistics?
    a. ____

2. The second thing to consider is residency. Place of birth is not defined by the location of the hospital in which a woman gives birth but by her residence. This makes sense as some small counties have no birthing facilities and if one used the place of birth, they would have no births registered. So you will need to be sure the woman resides in Tampa if you wish to count a birth as a "Tampa birth".

Once you have your case definition set, go through the birth certificates 1-51 and determine how many of these are teenage births for women in Tampa. The answers to this part of the workbook will be provided at the end, but be sure to do the exercise as this type of question may be on the test and you will need to be able to answer it. Write the totals in the workbook.

U.S. STANDARD CERTIFICATE OF LIVE BIRTH

LOCAL FILE NO. | BIRTH NUMBER:

CHILD	1. CHILD'S NAME (First, Middle, Last, Suffix) Thomas Sawyer	2. TIME OF BIRTH 13:46	3. SEX M	4. DATE OF BIRTH (Mo/Day/Yr) 2/13/2011
	5. FACILITY NAME (If not institution, give street and number) Brandon Regional Hospital	6. CITY, TOWN, OR LOCATION OF BIRTH Brandon	7. COUNTY OF BIRTH Hillsborough	
MOTHER	8a. MOTHER'S CURRENT LEGAL NAME (First, Middle, Last, Suffix) Jaime Sawyer	8b. DATE OF BIRTH (Mo/Day/Yr) 1/09/1994		
	8c. MOTHER'S NAME PRIOR TO FIRST MARRIAGE (First, Middle, Last, Suffix) Jaime Sawyer	8d. BIRTHPLACE (State, Territory, or Foreign Country) Missouri		
	9a. RESIDENCE OF MOTHER-STATE Florida	9b. COUNTY Hillsborough	9c. CITY, TOWN, OR LOCATION Tampa	
	9d. STREET AND NUMBER 4200 Fowler Avenue	9e. APT. NO. 115	9f. ZIP CODE 33620	9g. INSIDE CITY LIMITS? ■ Yes □ No

YES □ NO □

U.S. STANDARD CERTIFICATE OF LIVE BIRTH

LOCAL FILE NO. | BIRTH NUMBER:

CHILD	1. CHILD'S NAME (First, Middle, Last, Suffix) Jennifer Lopez	2. TIME OF BIRTH 6:42	3. SEX F	4. DATE OF BIRTH (Mo/Day/Yr) 4/19/2011
	5. FACILITY NAME (If not institution, give street and number) Tampa General Hospital	6. CITY, TOWN, OR LOCATION OF BIRTH Tampa	7. COUNTY OF BIRTH Hilllsborough	
MOTHER	8a. MOTHER'S CURRENT LEGAL NAME (First, Middle, Last, Suffix) Allison Lopez	8b. DATE OF BIRTH (Mo/Day/Yr) 1/28/1996		
	8c. MOTHER'S NAME PRIOR TO FIRST MARRIAGE (First, Middle, Last, Suffix) Allison Jimenez	8d. BIRTHPLACE (State, Territory, or Foreign Country) Florida		
	9a. RESIDENCE OF MOTHER-STATE Florida	9b. COUNTY Hillsborough	9c. CITY, TOWN, OR LOCATION Tampa	
	9d. STREET AND NUMBER 246 Fletcher Avenue	9e. APT. NO. -	9f. ZIP CODE 33617	9g. INSIDE CITY LIMITS? ■ Yes □ No

YES □ NO □

U.S. STANDARD CERTIFICATE OF LIVE BIRTH

LOCAL FILE NO. | BIRTH NUMBER:

CHILD	1. CHILD'S NAME (First, Middle, Last, Suffix) Jackie Kennedy	2. TIME OF BIRTH 13:46	3. SEX F	4. DATE OF BIRTH (Mo/Day/Yr) 11/04/2011
	5. FACILITY NAME (If not institution, give street and number) Tampa General Hospital	6. CITY, TOWN, OR LOCATION OF BIRTH Tampa	7. COUNTY OF BIRTH Hillsborough	
MOTHER	8a. MOTHER'S CURRENT LEGAL NAME (First, Middle, Last, Suffix) Mary Kennedy	8b. DATE OF BIRTH (Mo/Day/Yr) 05/27/1995		
	8c. MOTHER'S NAME PRIOR TO FIRST MARRIAGE (First, Middle, Last, Suffix) Mary Fowler	8d. BIRTHPLACE (State, Territory, or Foreign Country) Florida		
	9a. RESIDENCE OF MOTHER-STATE Florida	9b. COUNTY Hillsborough	9c. CITY, TOWN, OR LOCATION Tampa	
	9d. STREET AND NUMBER 4500 Tampa Palms Avenue	9e. APT. NO. 208	9f. ZIP CODE 33618	9g. INSIDE CITY LIMITS? ■ Yes □ No

YES □ NO □

U.S. STANDARD CERTIFICATE OF LIVE BIRTH

LOCAL FILE NO. BIRTH NUMBER:

CHILD	1. CHILD'S NAME (First, Middle, Last, Suffix) Joshua Sanders	2. TIME OF BIRTH 01:51	3. SEX M	4. DATE OF BIRTH (Mo/Day/Yr) 12/17/2011
	5. FACILITY NAME (If not institution, give street and number) Tampa General Hospital	6. CITY, TOWN, OR LOCATION OF BIRTH Tampa		7. COUNTY OF BIRTH Hillsborough
MOTHER	8a. MOTHER'S CURRENT LEGAL NAME (First, Middle, Last, Suffix) Nancy Sanders	8b. DATE OF BIRTH (Mo/Day/Yr) 06/28/1994		
	8c. MOTHER'S NAME PRIOR TO FIRST MARRIAGE (First, Middle, Last, Suffix) Nancy Schultz	8d. BIRTHPLACE (State, Territory, or Foreign Country) Florida		
	9a. RESIDENCE OF MOTHER-STATE Florida	9b. COUNTY Hillsborough	9c. CITY, TOWN, OR LOCATION Tampa	
	9d. STREET AND NUMBER 4850 Oak Vine Drive	9e. APT. NO. -	9f. ZIP CODE 33626	9g. INSIDE CITY LIMITS? ■ Yes □ No

YES ☐ NO ☐

U.S. STANDARD CERTIFICATE OF LIVE BIRTH

LOCAL FILE NO. BIRTH NUMBER:

CHILD	1. CHILD'S NAME (First, Middle, Last, Suffix) Kristen Hutchins	2. TIME OF BIRTH 20:35	3. SEX F	4. DATE OF BIRTH (Mo/Day/Yr) 03/21/2011
	5. FACILITY NAME (If not institution, give street and number) Tampa General Hospital	6. CITY, TOWN, OR LOCATION OF BIRTH Tampa		7. COUNTY OF BIRTH Hillsborough
MOTHER	8a. MOTHER'S CURRENT LEGAL NAME (First, Middle, Last, Suffix) Janice Wilson	8b. DATE OF BIRTH (Mo/Day/Yr) 02/05/1991		
	8c. MOTHER'S NAME PRIOR TO FIRST MARRIAGE (First, Middle, Last, Suffix) Janice Wilson	8d. BIRTHPLACE (State, Territory, or Foreign Country) Florida		
	9a. RESIDENCE OF MOTHER-STATE Florida	9b. COUNTY Hillsborough	9c. CITY, TOWN, OR LOCATION Tampa	
	9d. STREET AND NUMBER 1135 Gates Drive	9e. APT. NO. -	9f. ZIP CODE 33601	9g. INSIDE CITY LIMITS? ■ Yes □ No

YES ☐ NO ☐

U.S. STANDARD CERTIFICATE OF LIVE BIRTH

LOCAL FILE NO. BIRTH NUMBER:

CHILD	1. CHILD'S NAME (First, Middle, Last, Suffix) Theresa Miller	2. TIME OF BIRTH 23:35	3. SEX F	4. DATE OF BIRTH (Mo/Day/Yr) 09/14/2011
	5. FACILITY NAME (If not institution, give street and number) Tampa General Hospital	6. CITY, TOWN, OR LOCATION OF BIRTH Tampa		7. COUNTY OF BIRTH Hillsborough
MOTHER	8a. MOTHER'S CURRENT LEGAL NAME (First, Middle, Last, Suffix) Brittany Ann Miller	8b. DATE OF BIRTH (Mo/Day/Yr) 03/19/1992		
	8c. MOTHER'S NAME PRIOR TO FIRST MARRIAGE (First, Middle, Last, Suffix) Brittany Ann Hudson	8d. BIRTHPLACE (State, Territory, or Foreign Country) Florida		
	9a. RESIDENCE OF MOTHER-STATE Florida	9b. COUNTY Hillsborough	9c. CITY, TOWN, OR LOCATION Tampa	
	9d. STREET AND NUMBER 547 Soaring Avenue	9e. APT. NO. 105	9f. ZIP CODE 33631	9g. INSIDE CITY LIMITS? ■ Yes □ No

YES ☐ NO ☐

U.S. STANDARD CERTIFICATE OF LIVE BIRTH

LOCAL FILE NO. — BIRTH NUMBER:

CHILD

1. CHILD'S NAME (First, Middle, Last, Suffix)	2. TIME OF BIRTH	3. SEX	4. DATE OF BIRTH (Mo/Day/Yr)
Mark Andrew Miller	17:31	M	07/16/2011

5. FACILITY NAME (If not institution, give street and number)	6. CITY, TOWN, OR LOCATION OF BIRTH	7. COUNTY OF BIRTH
Tampa General Hospital	Tampa	Hillsborough

MOTHER

8a. MOTHER'S CURRENT LEGAL NAME (First, Middle, Last, Suffix)	8b. DATE OF BIRTH (Mo/Day/Yr)
Heather Grace Miller	11/29/1994
8c. MOTHER'S NAME PRIOR TO FIRST MARRIAGE (First, Middle, Last, Suffix)	**8d. BIRTHPLACE (State, Territory, or Foreign Country)**
Heather Grace Warner	Florida

9a. RESIDENCE OF MOTHER-STATE	9b. COUNTY	9c. CITY, TOWN, OR LOCATION
Florida	Hillsborough	Tampa

9d. STREET AND NUMBER	9e. APT. NO.	9f. ZIP CODE	9g. INSIDE CITY LIMITS?
352 Gibson Avenue	104	33611	■ Yes □ No

YES ☐ NO ☐

U.S. STANDARD CERTIFICATE OF LIVE BIRTH

LOCAL FILE NO. — BIRTH NUMBER:

CHILD

1. CHILD'S NAME (First, Middle, Last, Suffix)	2. TIME OF BIRTH	3. SEX	4. DATE OF BIRTH (Mo/Day/Yr)
Sarah Palin	14:33	F	04/30/2011

5. FACILITY NAME (If not institution, give street and number)	6. CITY, TOWN, OR LOCATION OF BIRTH	7. COUNTY OF BIRTH
University Community Hospital	Tampa	Hillsborough

MOTHER

8a. MOTHER'S CURRENT LEGAL NAME (First, Middle, Last, Suffix)	8b. DATE OF BIRTH (Mo/Day/Yr)
Brenda Ann Palin	11/13/1998
8c. MOTHER'S NAME PRIOR TO FIRST MARRIAGE (First, Middle, Last, Suffix)	**8d. BIRTHPLACE (State, Territory, or Foreign Country)**
Brenda Ann Palin	Canada

9a. RESIDENCE OF MOTHER-STATE	9b. COUNTY	9c. CITY, TOWN, OR LOCATION
Florida	Hillsborough	Wesley Chapel

9d. STREET AND NUMBER	9e. APT. NO.	9f. ZIP CODE	9g. INSIDE CITY LIMITS?
1536 Pine Point Drive	101	33659	■ Yes □ No

YES ☐ NO ☐

U.S. STANDARD CERTIFICATE OF LIVE BIRTH

LOCAL FILE NO. — BIRTH NUMBER:

CHILD

1. CHILD'S NAME (First, Middle, Last, Suffix)	2. TIME OF BIRTH	3. SEX	4. DATE OF BIRTH (Mo/Day/Yr)
John Snow	20:10	M	01/21/2011

5. FACILITY NAME (If not institution, give street and number)	6. CITY, TOWN, OR LOCATION OF BIRTH	7. COUNTY OF BIRTH
Tampa General Hospital	Tampa	Hillsborough

MOTHER

8a. MOTHER'S CURRENT LEGAL NAME (First, Middle, Last, Suffix)	8b. DATE OF BIRTH (Mo/Day/Yr)
Alicia Manning	12/12/1994
8c. MOTHER'S NAME PRIOR TO FIRST MARRIAGE (First, Middle, Last, Suffix)	**8d. BIRTHPLACE (State, Territory, or Foreign Country)**
Alicia Manning	Florida

9a. RESIDENCE OF MOTHER-STATE	9b. COUNTY	9c. CITY, TOWN, OR LOCATION
Florida	Hillsborough	Tampa

9d. STREET AND NUMBER	9e. APT. NO.	9f. ZIP CODE	9g. INSIDE CITY LIMITS?
837 Autumn Woods Avenue	34	33617	■ Yes □ No

YES ☐ NO ☐

U.S. STANDARD CERTIFICATE OF LIVE BIRTH

LOCAL FILE NO. | BIRTH NUMBER:

CHILD	1. CHILD'S NAME (First, Middle, Last, Suffix)	2. TIME OF BIRTH	3. SEX	4. DATE OF BIRTH (Mo/Day/Yr)
	Mary Jane Parker	13:21	F	04/12/2011

5. FACILITY NAME (If not institution, give street and number)	6. CITY, TOWN, OR LOCATION OF BIRTH	7. COUNTY OF BIRTH
Tampa General Hospital	Tampa	Hillsborough

MOTHER		
8a. MOTHER'S CURRENT LEGAL NAME (First, Middle, Last, Suffix)	8b. DATE OF BIRTH (Mo/Day/Yr)	
Alexis Parker	06/13/1994	
8c. MOTHER'S NAME PRIOR TO FIRST MARRIAGE (First, Middle, Last, Suffix)	8d. BIRTHPLACE (State, Territory, or Foreign Country)	
Alexis Collins	Florida	
9a. RESIDENCE OF MOTHER-STATE	9b. COUNTY	9c. CITY, TOWN, OR LOCATION
Florida	Hillsborough	Tampa

9d. STREET AND NUMBER	9e. APT. NO.	9f. ZIP CODE	9g. INSIDE CITY LIMITS?
3945 Regal Oaks Way	-	33685	■ Yes □ No

YES ☐ NO ☐

U.S. STANDARD CERTIFICATE OF LIVE BIRTH

LOCAL FILE NO. | BIRTH NUMBER:

CHILD	1. CHILD'S NAME (First, Middle, Last, Suffix)	2. TIME OF BIRTH	3. SEX	4. DATE OF BIRTH (Mo/Day/Yr)
	Shawn Knight	09:14	M	09/15/2011

5. FACILITY NAME (If not institution, give street and number)	6. CITY, TOWN, OR LOCATION OF BIRTH	7. COUNTY OF BIRTH
Tampa General Hospital	Tampa	HIllsborough

MOTHER		
8a. MOTHER'S CURRENT LEGAL NAME (First, Middle, Last, Suffix)	8b. DATE OF BIRTH (Mo/Day/Yr)	
Joanna Rose Knight	10/21/1991	
8c. MOTHER'S NAME PRIOR TO FIRST MARRIAGE (First, Middle, Last, Suffix)	8d. BIRTHPLACE (State, Territory, or Foreign Country)	
Joanna Rose Hill	Florida	
9a. RESIDENCE OF MOTHER-STATE	9b. COUNTY	9c. CITY, TOWN, OR LOCATION
Florida	HIllsborough	Tampa

9d. STREET AND NUMBER	9e. APT. NO.	9f. ZIP CODE	9g. INSIDE CITY LIMITS?
492 Sinclair Hills Road	-	33612	■ Yes □ No

YES ☐ NO ☐

U.S. STANDARD CERTIFICATE OF LIVE BIRTH

LOCAL FILE NO. | BIRTH NUMBER:

CHILD	1. CHILD'S NAME (First, Middle, Last, Suffix)	2. TIME OF BIRTH	3. SEX	4. DATE OF BIRTH (Mo/Day/Yr)
	Jennifer Aniston	15:45	F	08/22/2011

5. FACILITY NAME (If not institution, give street and number)	6. CITY, TOWN, OR LOCATION OF BIRTH	7. COUNTY OF BIRTH
Tampa General Hospital	Tampa	Hillsborough

MOTHER		
8a. MOTHER'S CURRENT LEGAL NAME (First, Middle, Last, Suffix)	8b. DATE OF BIRTH (Mo/Day/Yr)	
Kimberly Aniston	01/09/1994	
8c. MOTHER'S NAME PRIOR TO FIRST MARRIAGE (First, Middle, Last, Suffix)	8d. BIRTHPLACE (State, Territory, or Foreign Country)	
Kimberly Stocker	Florida	
9a. RESIDENCE OF MOTHER-STATE	9b. COUNTY	9c. CITY, TOWN, OR LOCATION
Florida	Hillsborough	Tampa

9d. STREET AND NUMBER	9e. APT. NO.	9f. ZIP CODE	9g. INSIDE CITY LIMITS?
112 North 56th Street	301	33610	■ Yes □ No

YES ☐ NO ☐

U.S. STANDARD CERTIFICATE OF LIVE BIRTH

LOCAL FILE NO. | BIRTH NUMBER:

CHILD

1. CHILD'S NAME (First, Middle, Last, Suffix)	2. TIME OF BIRTH	3. SEX	4. DATE OF BIRTH (Mo/Day/Yr)
Collin James Hume	21:05	M	09/07/2011

5. FACILITY NAME (If not institution, give street and number)	6. CITY, TOWN, OR LOCATION OF BIRTH	7. COUNTY OF BIRTH
University Community Hospital	Tampa	Hillsborough

MOTHER

8a. MOTHER'S CURRENT LEGAL NAME (First, Middle, Last, Suffix)	8b. DATE OF BIRTH (Mo/Day/Yr)
Brittany Evelyn Hume	05/19/1995
8c. MOTHER'S NAME PRIOR TO FIRST MARRIAGE (First, Middle, Last, Suffix)	**8d. BIRTHPLACE (State, Territory, or Foreign Country)**
Brittany Evelyn Raynor	Florida

9a. RESIDENCE OF MOTHER-STATE	9b. COUNTY	9c. CITY, TOWN, OR LOCATION
Florida	Hillsborough	Tampa

9d. STREET AND NUMBER	9e. APT. NO.	9f. ZIP CODE	9g. INSIDE CITY LIMITS?
583 Morning Drive	209	33623	■ Yes □ No

YES ☐ NO ☐

U.S. STANDARD CERTIFICATE OF LIVE BIRTH

LOCAL FILE NO. | BIRTH NUMBER:

CHILD

1. CHILD'S NAME (First, Middle, Last, Suffix)	2. TIME OF BIRTH	3. SEX	4. DATE OF BIRTH (Mo/Day/Yr)
Tiffany Renee Pierce	03:31	F	03/14/2011

5. FACILITY NAME (If not institution, give street and number)	6. CITY, TOWN, OR LOCATION OF BIRTH	7. COUNTY OF BIRTH
University Community Hospital	Tampa	Hillsborough

MOTHER

8a. MOTHER'S CURRENT LEGAL NAME (First, Middle, Last, Suffix)	8b. DATE OF BIRTH (Mo/Day/Yr)
Naomi Paige Pierce	09/17/1993
8c. MOTHER'S NAME PRIOR TO FIRST MARRIAGE (First, Middle, Last, Suffix)	**8d. BIRTHPLACE (State, Territory, or Foreign Country)**
Naomi Paige Kraft	Florida

9a. RESIDENCE OF MOTHER-STATE	9b. COUNTY	9c. CITY, TOWN, OR LOCATION
Florida	Hillsborough	Tampa

9d. STREET AND NUMBER	9e. APT. NO.	9f. ZIP CODE	9g. INSIDE CITY LIMITS?
115 Irene Street	-	33613	■ Yes □ No

YES ☐ NO ☐

U.S. STANDARD CERTIFICATE OF LIVE BIRTH

LOCAL FILE NO. | BIRTH NUMBER:

CHILD

1. CHILD'S NAME (First, Middle, Last, Suffix)	2. TIME OF BIRTH	3. SEX	4. DATE OF BIRTH (Mo/Day/Yr)
Rachel Autumn Lawton	02:05	F	12/25/2011

5. FACILITY NAME (If not institution, give street and number)	6. CITY, TOWN, OR LOCATION OF BIRTH	7. COUNTY OF BIRTH
Tampa General Hospital	Tampa	Hillsborough

MOTHER

8a. MOTHER'S CURRENT LEGAL NAME (First, Middle, Last, Suffix)	8b. DATE OF BIRTH (Mo/Day/Yr)
Holly May Lawton	07/08/1994
8c. MOTHER'S NAME PRIOR TO FIRST MARRIAGE (First, Middle, Last, Suffix)	**8d. BIRTHPLACE (State, Territory, or Foreign Country)**
Holly May Ritchey	Florida

9a. RESIDENCE OF MOTHER-STATE	9b. COUNTY	9c. CITY, TOWN, OR LOCATION
Florida	Hillsborough	Tampa

9d. STREET AND NUMBER	9e. APT. NO.	9f. ZIP CODE	9g. INSIDE CITY LIMITS?
3459 Liberty Avenue	101	33664	■ Yes □ No

YES ☐ NO ☐

U.S. STANDARD CERTIFICATE OF LIVE BIRTH

LOCAL FILE NO. BIRTH NUMBER:

CHILD	1. CHILD'S NAME (First, Middle, Last, Suffix)	2. TIME OF BIRTH	3. SEX	4. DATE OF BIRTH (Mo/Day/Yr)
	Alex Green	02:15	M	02/28/2011
	5. FACILITY NAME (If not institution, give street and number): St. Joseph's Women's Hospital	6. CITY, TOWN, OR LOCATION OF BIRTH: Tampa	7. COUNTY OF BIRTH: Hillsborough	
MOTHER	8a. MOTHER'S CURRENT LEGAL NAME (First, Middle, Last, Suffix): Jillian Elaine Green	8b. DATE OF BIRTH (Mo/Day/Yr): 11/02/1997		
	8c. MOTHER'S NAME PRIOR TO FIRST MARRIAGE (First, Middle, Last, Suffix): Jillian Elaine Taylor	8d. BIRTHPLACE (State, Territory, or Foreign Country): Florida		
	9a. RESIDENCE OF MOTHER-STATE: Florida	9b. COUNTY: Hillsborough	9c. CITY, TOWN, OR LOCATION: Tampa	
	9d. STREET AND NUMBER: 156 Forest Park Avenue	9e. APT. NO.: 106	9f. ZIP CODE: 33611	9g. INSIDE CITY LIMITS? ■ Yes □ No

YES ☐ NO ☐

U.S. STANDARD CERTIFICATE OF LIVE BIRTH

LOCAL FILE NO. BIRTH NUMBER:

CHILD	1. CHILD'S NAME (First, Middle, Last, Suffix)	2. TIME OF BIRTH	3. SEX	4. DATE OF BIRTH (Mo/Day/Yr)
	Abigail Chan	02:19	F	09/11/2011
	5. FACILITY NAME (If not institution, give street and number): Tampa General Hospital	6. CITY, TOWN, OR LOCATION OF BIRTH: Tampa	7. COUNTY OF BIRTH: Hillsborough	
MOTHER	8a. MOTHER'S CURRENT LEGAL NAME (First, Middle, Last, Suffix): Rebecca Lauren Chan	8b. DATE OF BIRTH (Mo/Day/Yr): 09/01/1996		
	8c. MOTHER'S NAME PRIOR TO FIRST MARRIAGE (First, Middle, Last, Suffix): Rebecca Lauren Richardson	8d. BIRTHPLACE (State, Territory, or Foreign Country): Florida		
	9a. RESIDENCE OF MOTHER-STATE: Florida	9b. COUNTY: Hillsborough	9c. CITY, TOWN, OR LOCATION: Tampa	
	9d. STREET AND NUMBER: 1195 Nantucket Drive	9e. APT. NO.: -	9f. ZIP CODE: 33631	9g. INSIDE CITY LIMITS? ■ Yes □ No

YES ☐ NO ☐

U.S. STANDARD CERTIFICATE OF LIVE BIRTH

LOCAL FILE NO. BIRTH NUMBER:

CHILD	1. CHILD'S NAME (First, Middle, Last, Suffix)	2. TIME OF BIRTH	3. SEX	4. DATE OF BIRTH (Mo/Day/Yr)
	Philip Tyler Park	11:59	M	11/11/2011
	5. FACILITY NAME (If not institution, give street and number): Tampa General Hospital	6. CITY, TOWN, OR LOCATION OF BIRTH: Tampa	7. COUNTY OF BIRTH: Hillsborough	
MOTHER	8a. MOTHER'S CURRENT LEGAL NAME (First, Middle, Last, Suffix): Erin Olivia Park	8b. DATE OF BIRTH (Mo/Day/Yr): 11/13/1995		
	8c. MOTHER'S NAME PRIOR TO FIRST MARRIAGE (First, Middle, Last, Suffix): Erin Olivia Suarez	8d. BIRTHPLACE (State, Territory, or Foreign Country): Florida		
	9a. RESIDENCE OF MOTHER-STATE: Florida	9b. COUNTY: HIllsborough	9c. CITY, TOWN, OR LOCATION: Tampa	
	9d. STREET AND NUMBER: 1956 Cliff Drive	9e. APT. NO.: 107	9f. ZIP CODE: 33625	9g. INSIDE CITY LIMITS? ■ Yes □ No

YES ☐ NO ☐

U.S. STANDARD CERTIFICATE OF LIVE BIRTH

LOCAL FILE NO. BIRTH NUMBER:

CHILD

1. CHILD'S NAME (First, Middle, Last, Suffix)	2. TIME OF BIRTH	3. SEX	4. DATE OF BIRTH (Mo/Day/Yr)
Stephen Colbert, Jr.	12:56	M	06/08/2011

5. FACILITY NAME (If not institution, give street and number)	6. CITY, TOWN, OR LOCATION OF BIRTH	7. COUNTY OF BIRTH
Tampa General Hospital	Tampa	Hillsborough

MOTHER

8a. MOTHER'S CURRENT LEGAL NAME (First, Middle, Last, Suffix)	8b. DATE OF BIRTH (Mo/Day/Yr)
Bonnie Kara Carter	11/22/1997
8c. MOTHER'S NAME PRIOR TO FIRST MARRIAGE (First, Middle, Last, Suffix)	8d. BIRTHPLACE (State, Territory, or Foreign Country)
Bonnie Kara Carter	Florida

9a. RESIDENCE OF MOTHER-STATE	9b. COUNTY	9c. CITY, TOWN, OR LOCATION
Florida	Hillsborough	Tampa

9d. STREET AND NUMBER	9e. APT. NO.	9f. ZIP CODE	9g. INSIDE CITY LIMITS?
567 Myrtle Street	406	33616	■ Yes □ No

YES □ NO □

U.S. STANDARD CERTIFICATE OF LIVE BIRTH

LOCAL FILE NO. BIRTH NUMBER:

CHILD

1. CHILD'S NAME (First, Middle, Last, Suffix)	2. TIME OF BIRTH	3. SEX	4. DATE OF BIRTH (Mo/Day/Yr)
Charles Travis Schultz	18:04	M	06/25/2011

5. FACILITY NAME (If not institution, give street and number)	6. CITY, TOWN, OR LOCATION OF BIRTH	7. COUNTY OF BIRTH
Tampa General Hospital	Tampa	Hillsborough

MOTHER

8a. MOTHER'S CURRENT LEGAL NAME (First, Middle, Last, Suffix)	8b. DATE OF BIRTH (Mo/Day/Yr)
Stephanie Allison Schultz	01/09/1994
8c. MOTHER'S NAME PRIOR TO FIRST MARRIAGE (First, Middle, Last, Suffix)	8d. BIRTHPLACE (State, Territory, or Foreign Country)
Stephanie Allison Sheldon	Florida

9a. RESIDENCE OF MOTHER-STATE	9b. COUNTY	9c. CITY, TOWN, OR LOCATION
Florida	Hillsborough	Brandon

9d. STREET AND NUMBER	9e. APT. NO.	9f. ZIP CODE	9g. INSIDE CITY LIMITS?
225 Parade Street	-	33508	■ Yes □ No

YES □ NO □

U.S. STANDARD CERTIFICATE OF LIVE BIRTH

LOCAL FILE NO. BIRTH NUMBER:

CHILD

1. CHILD'S NAME (First, Middle, Last, Suffix)	2. TIME OF BIRTH	3. SEX	4. DATE OF BIRTH (Mo/Day/Yr)
Desiree Leah Getter	22:15	F	06/27/2011

5. FACILITY NAME (If not institution, give street and number)	6. CITY, TOWN, OR LOCATION OF BIRTH	7. COUNTY OF BIRTH
Tampa General Hospital	Tampa	Hillsborough

MOTHER

8a. MOTHER'S CURRENT LEGAL NAME (First, Middle, Last, Suffix)	8b. DATE OF BIRTH (Mo/Day/Yr)
Katherine Tricia Getter	12/25/1993
8c. MOTHER'S NAME PRIOR TO FIRST MARRIAGE (First, Middle, Last, Suffix)	8d. BIRTHPLACE (State, Territory, or Foreign Country)
Katherine Tricia Getter	Florida

9a. RESIDENCE OF MOTHER-STATE	9b. COUNTY	9c. CITY, TOWN, OR LOCATION
Florida	Hillsborough	Tampa

9d. STREET AND NUMBER	9e. APT. NO.	9f. ZIP CODE	9g. INSIDE CITY LIMITS?
1536 Matador Court	301	33611	■ Yes □ No

YES □ NO □

U.S. STANDARD CERTIFICATE OF LIVE BIRTH

LOCAL FILE NO. | BIRTH NUMBER:

CHILD	1. CHILD'S NAME (First, Middle, Last, Suffix) Michael James Jordan	2. TIME OF BIRTH 13:01	3. SEX M	4. DATE OF BIRTH (Mo/Day/Yr) 4/15/2011
	5. FACILITY NAME (If not institution, give street and number) Tampa General Hospital	6. CITY, TOWN, OR LOCATION OF BIRTH Tampa	7. COUNTY OF BIRTH Hillsborough	
MOTHER	8a. MOTHER'S CURRENT LEGAL NAME (First, Middle, Last, Suffix) Shannon Jordan	8b. DATE OF BIRTH (Mo/Day/Yr) 03/28/1992		
	8c. MOTHER'S NAME PRIOR TO FIRST MARRIAGE (First, Middle, Last, Suffix) Shannon Holtz	8d. BIRTHPLACE (State, Territory, or Foreign Country) Florida		
	9a. RESIDENCE OF MOTHER-STATE Florida	9b. COUNTY Hillsborough	9c. CITY, TOWN, OR LOCATION Tampa	
	9d. STREET AND NUMBER 2351 Overlook Drive	9e. APT. NO. -	9f. ZIP CODE 33603	9g. INSIDE CITY LIMITS? ■ Yes □ No

YES ☐ NO ☐

U.S. STANDARD CERTIFICATE OF LIVE BIRTH

LOCAL FILE NO. | BIRTH NUMBER:

CHILD	1. CHILD'S NAME (First, Middle, Last, Suffix) Amelia Jane Mercer	2. TIME OF BIRTH 05:36	3. SEX F	4. DATE OF BIRTH (Mo/Day/Yr) 07/30/2011
	5. FACILITY NAME (If not institution, give street and number) Tampa General Hospital	6. CITY, TOWN, OR LOCATION OF BIRTH Tampa	7. COUNTY OF BIRTH Hillsborough	
MOTHER	8a. MOTHER'S CURRENT LEGAL NAME (First, Middle, Last, Suffix) Christine Amber Mercer	8b. DATE OF BIRTH (Mo/Day/Yr) 04/12/1994		
	8c. MOTHER'S NAME PRIOR TO FIRST MARRIAGE (First, Middle, Last, Suffix) Christine Amber Talbert	8d. BIRTHPLACE (State, Territory, or Foreign Country) Florida		
	9a. RESIDENCE OF MOTHER-STATE Florida	9b. COUNTY Hillsborough	9c. CITY, TOWN, OR LOCATION Tampa	
	9d. STREET AND NUMBER 235 Regnas Avenue	9e. APT. NO. 3	9f. ZIP CODE 33630	9g. INSIDE CITY LIMITS? ■ Yes □ No

YES ☐ NO ☐

U.S. STANDARD CERTIFICATE OF LIVE BIRTH

LOCAL FILE NO. | BIRTH NUMBER:

CHILD	1. CHILD'S NAME (First, Middle, Last, Suffix) Nicole Gail Garrett	2. TIME OF BIRTH 16:25	3. SEX F	4. DATE OF BIRTH (Mo/Day/Yr) 03/01/2011
	5. FACILITY NAME (If not institution, give street and number) Cape Canaveral Hospital	6. CITY, TOWN, OR LOCATION OF BIRTH Cocoa Beach	7. COUNTY OF BIRTH Brevard	
MOTHER	8a. MOTHER'S CURRENT LEGAL NAME (First, Middle, Last, Suffix) Jeanette Garrett	8b. DATE OF BIRTH (Mo/Day/Yr) 07/04/1991		
	8c. MOTHER'S NAME PRIOR TO FIRST MARRIAGE (First, Middle, Last, Suffix) Jeanette Garrett	8d. BIRTHPLACE (State, Territory, or Foreign Country) Arizona		
	9a. RESIDENCE OF MOTHER-STATE Florida	9b. COUNTY Brevard	9c. CITY, TOWN, OR LOCATION Cocoa Beach	
	9d. STREET AND NUMBER 701 North 1st Street	9e. APT. NO. -	9f. ZIP CODE 32931	9g. INSIDE CITY LIMITS? ■ Yes □ No

YES ☐ NO ☐

U.S. STANDARD CERTIFICATE OF LIVE BIRTH

LOCAL FILE NO.

BIRTH NUMBER:

CHILD	1. CHILD'S NAME (First, Middle, Last, Suffix): Marissa May Markley	2. TIME OF BIRTH: 10:16	3. SEX: F	4. DATE OF BIRTH (Mo/Day/Yr): 06/26/2011
	5. FACILITY NAME (If not institution, give street and number): Tampa General Hospital	6. CITY, TOWN, OR LOCATION OF BIRTH: Tampa	7. COUNTY OF BIRTH: Hillsborough	
MOTHER	8a. MOTHER'S CURRENT LEGAL NAME (First, Middle, Last, Suffix): Samantha Markley	8b. DATE OF BIRTH (Mo/Day/Yr): 01/30/1997		
	8c. MOTHER'S NAME PRIOR TO FIRST MARRIAGE (First, Middle, Last, Suffix): Samantha Rice	8d. BIRTHPLACE (State, Territory, or Foreign Country): Florida		
	9a. RESIDENCE OF MOTHER-STATE: Florida	9b. COUNTY: Hillsborough	9c. CITY, TOWN, OR LOCATION: Tampa	
	9d. STREET AND NUMBER: 532 Brentwood Drive	9e. APT. NO.: -	9f. ZIP CODE: 33618	9g. INSIDE CITY LIMITS? ■ Yes □ No

YES ☐ NO ☐

U.S. STANDARD CERTIFICATE OF LIVE BIRTH

LOCAL FILE NO.

BIRTH NUMBER:

CHILD	1. CHILD'S NAME (First, Middle, Last, Suffix): Kobe Bryant	2. TIME OF BIRTH: 13:55	3. SEX: M	4. DATE OF BIRTH (Mo/Day/Yr): 06/26/2011
	5. FACILITY NAME (If not institution, give street and number): Tampa General Hospital	6. CITY, TOWN, OR LOCATION OF BIRTH: Tampa	7. COUNTY OF BIRTH: Hillsborough	
MOTHER	8a. MOTHER'S CURRENT LEGAL NAME (First, Middle, Last, Suffix): Beatrice Bryant	8b. DATE OF BIRTH (Mo/Day/Yr): 12/15/1993		
	8c. MOTHER'S NAME PRIOR TO FIRST MARRIAGE (First, Middle, Last, Suffix): Beatrice Bryant	8d. BIRTHPLACE (State, Territory, or Foreign Country): Florida		
	9a. RESIDENCE OF MOTHER-STATE: Florida	9b. COUNTY: Hillsborough	9c. CITY, TOWN, OR LOCATION: Tampa	
	9d. STREET AND NUMBER: 313 Pine Hill Drive	9e. APT. NO.: -	9f. ZIP CODE: 33624	9g. INSIDE CITY LIMITS? ■ Yes □ No

YES ☐ NO ☐

U.S. STANDARD CERTIFICATE OF LIVE BIRTH

LOCAL FILE NO.

BIRTH NUMBER:

CHILD	1. CHILD'S NAME (First, Middle, Last, Suffix): Julie Andrews	2. TIME OF BIRTH: 21:15	3. SEX: F	4. DATE OF BIRTH (Mo/Day/Yr): 05/16/2011
	5. FACILITY NAME (If not institution, give street and number): Tampa General Hospital	6. CITY, TOWN, OR LOCATION OF BIRTH: Tampa	7. COUNTY OF BIRTH: Hillsborough	
MOTHER	8a. MOTHER'S CURRENT LEGAL NAME (First, Middle, Last, Suffix): Hannah Andrews	8b. DATE OF BIRTH (Mo/Day/Yr): 03/05/1991		
	8c. MOTHER'S NAME PRIOR TO FIRST MARRIAGE (First, Middle, Last, Suffix): Hannah Andrews	8d. BIRTHPLACE (State, Territory, or Foreign Country): Florida		
	9a. RESIDENCE OF MOTHER-STATE: Florida	9b. COUNTY: Hillsborough	9c. CITY, TOWN, OR LOCATION: Tampa	
	9d. STREET AND NUMBER: 1156 Flounder Avenue	9e. APT. NO.: -	9f. ZIP CODE: 33623	9g. INSIDE CITY LIMITS? ■ Yes □ No

YES ☐ NO ☐

U.S. STANDARD CERTIFICATE OF LIVE BIRTH

LOCAL FILE NO. BIRTH NUMBER:

CHILD	1. CHILD'S NAME (First, Middle, Last, Suffix)	2. TIME OF BIRTH	3. SEX	4. DATE OF BIRTH (Mo/Day/Yr)
	Evelyn Bailey Spears	13:55	F	10/30/2011
	5. FACILITY NAME (If not institution, give street and number)	6. CITY, TOWN, OR LOCATION OF BIRTH	7. COUNTY OF BIRTH	
	Tampa General Hospital	Tampa	Hillsborough	
MOTHER	8a. MOTHER'S CURRENT LEGAL NAME (First, Middle, Last, Suffix)	8b. DATE OF BIRTH (Mo/Day/Yr)		
	Brittany Autumn Spears	06/23/1992		
	8c. MOTHER'S NAME PRIOR TO FIRST MARRIAGE (First, Middle, Last, Suffix)	8d. BIRTHPLACE (State, Territory, or Foreign Country)		
	Brittany Autumn Oakley	Florida		
	9a. RESIDENCE OF MOTHER-STATE	9b. COUNTY	9c. CITY, TOWN, OR LOCATION	
	Florida	Hillsborough	Tampa	
	9d. STREET AND NUMBER	9e. APT. NO.	9f. ZIP CODE	9g. INSIDE CITY LIMITS?
	3352 Shadow Lane	-	33629	■ Yes □ No

YES ☐ NO ☐

U.S. STANDARD CERTIFICATE OF LIVE BIRTH

LOCAL FILE NO. BIRTH NUMBER:

CHILD	1. CHILD'S NAME (First, Middle, Last, Suffix)	2. TIME OF BIRTH	3. SEX	4. DATE OF BIRTH (Mo/Day/Yr)
	Chloe Jackson	4:23	F	07/04/2011
	5. FACILITY NAME (If not institution, give street and number)	6. CITY, TOWN, OR LOCATION OF BIRTH	7. COUNTY OF BIRTH	
	Tampa General Hospital	Tampa	Hillsborough	
MOTHER	8a. MOTHER'S CURRENT LEGAL NAME (First, Middle, Last, Suffix)	8b. DATE OF BIRTH (Mo/Day/Yr)		
	Lillian Towers	12/16/1992		
	8c. MOTHER'S NAME PRIOR TO FIRST MARRIAGE (First, Middle, Last, Suffix)	8d. BIRTHPLACE (State, Territory, or Foreign Country)		
	Lillian Towers	Missouri		
	9a. RESIDENCE OF MOTHER-STATE	9b. COUNTY	9c. CITY, TOWN, OR LOCATION	
	Florida	Hillsborough	Tampa	
	9d. STREET AND NUMBER	9e. APT. NO.	9f. ZIP CODE	9g. INSIDE CITY LIMITS?
	193 Grove Hill Road	-	33674	■ Yes □ No

YES ☐ NO ☐

U.S. STANDARD CERTIFICATE OF LIVE BIRTH

LOCAL FILE NO. BIRTH NUMBER:

CHILD	1. CHILD'S NAME (First, Middle, Last, Suffix)	2. TIME OF BIRTH	3. SEX	4. DATE OF BIRTH (Mo/Day/Yr)
	Katherine Grace Gould	06:29	F	01/03/2011
	5. FACILITY NAME (If not institution, give street and number)	6. CITY, TOWN, OR LOCATION OF BIRTH	7. COUNTY OF BIRTH	
	Tampa General Hospital	Tampa	Hillsborough	
MOTHER	8a. MOTHER'S CURRENT LEGAL NAME (First, Middle, Last, Suffix)	8b. DATE OF BIRTH (Mo/Day/Yr)		
	Vanessa Harris	06/13/1991		
	8c. MOTHER'S NAME PRIOR TO FIRST MARRIAGE (First, Middle, Last, Suffix)	8d. BIRTHPLACE (State, Territory, or Foreign Country)		
	Vanessa Walker	Florida		
	9a. RESIDENCE OF MOTHER-STATE	9b. COUNTY	9c. CITY, TOWN, OR LOCATION	
	Florida	Hillsborough	Tampa	
	9d. STREET AND NUMBER	9e. APT. NO.	9f. ZIP CODE	9g. INSIDE CITY LIMITS?
	335 Ridgeway Road	-	33684	■ Yes □ No

YES ☐ NO ☐

U.S. STANDARD CERTIFICATE OF LIVE BIRTH

LOCAL FILE NO. BIRTH NUMBER:

CHILD	1. CHILD'S NAME (First, Middle, Last, Suffix)	2. TIME OF BIRTH	3. SEX	4. DATE OF BIRTH (Mo/Day/Yr)
	Brianna Miller	08:33	F	10/11/2011
	5. FACILITY NAME (If not institution, give street and number): Tampa General Hospital	6. CITY, TOWN, OR LOCATION OF BIRTH: Tampa		7. COUNTY OF BIRTH: Hillsborough
MOTHER	8a. MOTHER'S CURRENT LEGAL NAME (First, Middle, Last, Suffix): Melanie Neilsen	8b. DATE OF BIRTH (Mo/Day/Yr): 11/17/1997		
	8c. MOTHER'S NAME PRIOR TO FIRST MARRIAGE (First, Middle, Last, Suffix): Melanie Miller	8d. BIRTHPLACE (State, Territory, or Foreign Country): California		
	9a. RESIDENCE OF MOTHER-STATE: Florida	9b. COUNTY: Hillsborough	9c. CITY, TOWN, OR LOCATION: Tampa	
	9d. STREET AND NUMBER: 2252 River Dune Street	9e. APT. NO.: -	9f. ZIP CODE: 33615	9g. INSIDE CITY LIMITS? ■ Yes □ No

YES ☐ NO ☐

U.S. STANDARD CERTIFICATE OF LIVE BIRTH

LOCAL FILE NO. BIRTH NUMBER:

CHILD	1. CHILD'S NAME (First, Middle, Last, Suffix)	2. TIME OF BIRTH	3. SEX	4. DATE OF BIRTH (Mo/Day/Yr)
	Antonio Goodman	13:26	M	10/11/2011
	5. FACILITY NAME (If not institution, give street and number): Tampa General Hospital	6. CITY, TOWN, OR LOCATION OF BIRTH: Tampa		7. COUNTY OF BIRTH: Hillsborough
MOTHER	8a. MOTHER'S CURRENT LEGAL NAME (First, Middle, Last, Suffix): Olivia Goodman	8b. DATE OF BIRTH (Mo/Day/Yr): 03/03/1991		
	8c. MOTHER'S NAME PRIOR TO FIRST MARRIAGE (First, Middle, Last, Suffix): Olivia Evans	8d. BIRTHPLACE (State, Territory, or Foreign Country): Georgia		
	9a. RESIDENCE OF MOTHER-STATE: Florida	9b. COUNTY: Hillsborough	9c. CITY, TOWN, OR LOCATION: Tampa	
	9d. STREET AND NUMBER: 331 Orange Place	9e. APT. NO.: -	9f. ZIP CODE: 33613	9g. INSIDE CITY LIMITS? ■ Yes □ No

YES ☐ NO ☐

U.S. STANDARD CERTIFICATE OF LIVE BIRTH

LOCAL FILE NO. BIRTH NUMBER:

CHILD	1. CHILD'S NAME (First, Middle, Last, Suffix)	2. TIME OF BIRTH	3. SEX	4. DATE OF BIRTH (Mo/Day/Yr)
	Christopher Thomas Decker	17:01	M	07/26/2011
	5. FACILITY NAME (If not institution, give street and number): Tampa General Hospital	6. CITY, TOWN, OR LOCATION OF BIRTH: Tampa		7. COUNTY OF BIRTH: Hillsborough
MOTHER	8a. MOTHER'S CURRENT LEGAL NAME (First, Middle, Last, Suffix): Paige Green	8b. DATE OF BIRTH (Mo/Day/Yr): 05/04/1994		
	8c. MOTHER'S NAME PRIOR TO FIRST MARRIAGE (First, Middle, Last, Suffix): Paige Green	8d. BIRTHPLACE (State, Territory, or Foreign Country): Florida		
	9a. RESIDENCE OF MOTHER-STATE: Florida	9b. COUNTY: Hillsborough	9c. CITY, TOWN, OR LOCATION: Tampa	
	9d. STREET AND NUMBER: 945 Mango Terrace	9e. APT. NO.: -	9f. ZIP CODE: 33611	9g. INSIDE CITY LIMITS? ■ Yes □ No

YES ☐ NO ☐

U.S. STANDARD CERTIFICATE OF LIVE BIRTH					
LOCAL FILE NO.					BIRTH NUMBER:
CHILD	1. CHILD'S NAME (First, Middle, Last, Suffix) Addison Faith Hammock		2. TIME OF BIRTH 20:15	3. SEX F	4. DATE OF BIRTH (Mo/Day/Yr) 02/24/2011
	5. FACILITY NAME (If not institution, give street and number) Tampa General Hospital	6. CITY, TOWN, OR LOCATION OF BIRTH Tampa		7. COUNTY OF BIRTH Hillsborough	
MOTHER	8a. MOTHER'S CURRENT LEGAL NAME (First, Middle, Last, Suffix) Emma Watson		8b. DATE OF BIRTH (Mo/Day/Yr) 06/27/1993		
	8c. MOTHER'S NAME PRIOR TO FIRST MARRIAGE (First, Middle, Last, Suffix) Emma Watson		8d. BIRTHPLACE (State, Territory, or Foreign Country) Wisconsin		
	9a. RESIDENCE OF MOTHER-STATE Florida	9b. COUNTY Hillsborough	9c. CITY, TOWN, OR LOCATION Tampa		
	9d. STREET AND NUMBER 3331 Mission Hills Avenue	9e. APT. NO. -	9f. ZIP CODE 33687		9g. INSIDE CITY LIMITS? ■ Yes □ No

YES ☐ NO ☐

U.S. STANDARD CERTIFICATE OF LIVE BIRTH					
LOCAL FILE NO.					BIRTH NUMBER:
CHILD	1. CHILD'S NAME (First, Middle, Last, Suffix) Jocelyn Sandberg		2. TIME OF BIRTH 10:91	3. SEX F	4. DATE OF BIRTH (Mo/Day/Yr) 05/20/2011
	5. FACILITY NAME (If not institution, give street and number) Brandon Regional Hospital	6. CITY, TOWN, OR LOCATION OF BIRTH Brandon		7. COUNTY OF BIRTH Hillsborough	
MOTHER	8a. MOTHER'S CURRENT LEGAL NAME (First, Middle, Last, Suffix) Layla Hudson		8b. DATE OF BIRTH (Mo/Day/Yr) 04/12/1996		
	8c. MOTHER'S NAME PRIOR TO FIRST MARRIAGE (First, Middle, Last, Suffix) Layla Hudson		8d. BIRTHPLACE (State, Territory, or Foreign Country) Pennsylvania		
	9a. RESIDENCE OF MOTHER-STATE Florida	9b. COUNTY Hillsborough	9c. CITY, TOWN, OR LOCATION Tampa		
	9d. STREET AND NUMBER 315 Temple Heights road	9e. APT. NO. -	9f. ZIP CODE 33612		9g. INSIDE CITY LIMITS? ■ Yes □ No

YES ☐ NO ☐

U.S. STANDARD CERTIFICATE OF LIVE BIRTH					
LOCAL FILE NO.					BIRTH NUMBER:
CHILD	1. CHILD'S NAME (First, Middle, Last, Suffix) Chad West		2. TIME OF BIRTH 17:12	3. SEX M	4. DATE OF BIRTH (Mo/Day/Yr) 04/09/2011
	5. FACILITY NAME (If not institution, give street and number) Tampa General Hospital	6. CITY, TOWN, OR LOCATION OF BIRTH Tampa		7. COUNTY OF BIRTH Hillsborough	
MOTHER	8a. MOTHER'S CURRENT LEGAL NAME (First, Middle, Last, Suffix) Tiffany Thompson		8b. DATE OF BIRTH (Mo/Day/Yr) 03/30/1993		
	8c. MOTHER'S NAME PRIOR TO FIRST MARRIAGE (First, Middle, Last, Suffix) Tiffany Donahue		8d. BIRTHPLACE (State, Territory, or Foreign Country) Florida		
	9a. RESIDENCE OF MOTHER-STATE Florida	9b. COUNTY Hillsborough	9c. CITY, TOWN, OR LOCATION Tampa		
	9d. STREET AND NUMBER 139 Holland Street	9e. APT. NO. -	9f. ZIP CODE 33624		9g. INSIDE CITY LIMITS? ■ Yes □ No

YES ☐ NO ☐

U.S. STANDARD CERTIFICATE OF LIVE BIRTH

LOCAL FILE NO. | BIRTH NUMBER:

CHILD	1. CHILD'S NAME (First, Middle, Last, Suffix) Derek Davenport		2. TIME OF BIRTH 12:19	3. SEX M	4. DATE OF BIRTH (Mo/Day/Yr) 04/27/2011
	5. FACILITY NAME (If not institution, give street and number) Tampa General Hospital		6. CITY, TOWN, OR LOCATION OF BIRTH Tampa		7. COUNTY OF BIRTH Hillsborough
MOTHER	8a. MOTHER'S CURRENT LEGAL NAME (First, Middle, Last, Suffix) Cassie Davenport		8b. DATE OF BIRTH (Mo/Day/Yr) 05/20/1992		
	8c. MOTHER'S NAME PRIOR TO FIRST MARRIAGE (First, Middle, Last, Suffix) Cassie Davis		8d. BIRTHPLACE (State, Territory, or Foreign Country) Oregon		
	9a. RESIDENCE OF MOTHER-STATE Florida	9b. COUNTY Hillsborough	9c. CITY, TOWN, OR LOCATION Tampa		
	9d. STREET AND NUMBER 776 Peach Drive	9e. APT. NO. -	9f. ZIP CODE 33608		9g. INSIDE CITY LIMITS? ■ Yes □ No

YES ☐ NO ☐

U.S. STANDARD CERTIFICATE OF LIVE BIRTH

LOCAL FILE NO. | BIRTH NUMBER:

CHILD	1. CHILD'S NAME (First, Middle, Last, Suffix) Steven Carson		2. TIME OF BIRTH 18:22	3. SEX M	4. DATE OF BIRTH (Mo/Day/Yr) 08/18/2011
	5. FACILITY NAME (If not institution, give street and number) Tampa General Hospital		6. CITY, TOWN, OR LOCATION OF BIRTH Tampa		7. COUNTY OF BIRTH Hillsborough
MOTHER	8a. MOTHER'S CURRENT LEGAL NAME (First, Middle, Last, Suffix) Suzanne Baker		8b. DATE OF BIRTH (Mo/Day/Yr) 05/20/1991		
	8c. MOTHER'S NAME PRIOR TO FIRST MARRIAGE (First, Middle, Last, Suffix) Suzanne Baker		8d. BIRTHPLACE (State, Territory, or Foreign Country) Texas		
	9a. RESIDENCE OF MOTHER-STATE Florida	9b. COUNTY Hillsborough	9c. CITY, TOWN, OR LOCATION Tampa		
	9d. STREET AND NUMBER 837 Gulf Court	9e. APT. NO. -	9f. ZIP CODE 33684		9g. INSIDE CITY LIMITS? ■ Yes □ No

YES ☐ NO ☐

U.S. STANDARD CERTIFICATE OF LIVE BIRTH

LOCAL FILE NO. | BIRTH NUMBER:

CHILD	1. CHILD'S NAME (First, Middle, Last, Suffix) Matthew Grothe		2. TIME OF BIRTH 14:39	3. SEX M	4. DATE OF BIRTH (Mo/Day/Yr) 05/19/2011
	5. FACILITY NAME (If not institution, give street and number) Tampa General Hospital		6. CITY, TOWN, OR LOCATION OF BIRTH Tampa		7. COUNTY OF BIRTH Hillsborough
MOTHER	8a. MOTHER'S CURRENT LEGAL NAME (First, Middle, Last, Suffix) Samantha Grothe		8b. DATE OF BIRTH (Mo/Day/Yr) 04/17/1993		
	8c. MOTHER'S NAME PRIOR TO FIRST MARRIAGE (First, Middle, Last, Suffix) Samantha Murphy		8d. BIRTHPLACE (State, Territory, or Foreign Country) North Carolina		
	9a. RESIDENCE OF MOTHER-STATE Florida	9b. COUNTY Hillsborough	9c. CITY, TOWN, OR LOCATION Tampa		
	9d. STREET AND NUMBER 4202 East Fowler Avenue	9e. APT. NO. -	9f. ZIP CODE 33620		9g. INSIDE CITY LIMITS? ■ Yes □ No

YES ☐ NO ☐

U.S. STANDARD CERTIFICATE OF LIVE BIRTH

LOCAL FILE NO. | BIRTH NUMBER:

CHILD	1. CHILD'S NAME (First, Middle, Last, Suffix)	2. TIME OF BIRTH	3. SEX	4. DATE OF BIRTH (Mo/Day/Yr)
	Joseph Stetson	20:15	M	08/29/2011

5. FACILITY NAME (If not institution, give street and number)	6. CITY, TOWN, OR LOCATION OF BIRTH	7. COUNTY OF BIRTH
Tampa General Hospital	Tampa	Hillsborough

MOTHER	8a. MOTHER'S CURRENT LEGAL NAME (First, Middle, Last, Suffix)	8b. DATE OF BIRTH (Mo/Day/Yr)
	Kathy Harris	01/02/1993
	8c. MOTHER'S NAME PRIOR TO FIRST MARRIAGE (First, Middle, Last, Suffix)	8d. BIRTHPLACE (State, Territory, or Foreign Country)
	Kathy Harris	Rhode Island

9a. RESIDENCE OF MOTHER-STATE	9b. COUNTY	9c. CITY, TOWN, OR LOCATION
Florida	Hillsborough	Tampa

9d. STREET AND NUMBER	9e. APT. NO.	9f. ZIP CODE	9g. INSIDE CITY LIMITS?
1024 Aspen Avenue	-	33611	■ Yes □ No

YES □ NO □

U.S. STANDARD CERTIFICATE OF LIVE BIRTH

LOCAL FILE NO. | BIRTH NUMBER:

CHILD	1. CHILD'S NAME (First, Middle, Last, Suffix)	2. TIME OF BIRTH	3. SEX	4. DATE OF BIRTH (Mo/Day/Yr)
	Caroline Moore	11:33	F	04/08/2011

5. FACILITY NAME (If not institution, give street and number)	6. CITY, TOWN, OR LOCATION OF BIRTH	7. COUNTY OF BIRTH
Tampa General Hospital	Tampa	Hillsborough

MOTHER	8a. MOTHER'S CURRENT LEGAL NAME (First, Middle, Last, Suffix)	8b. DATE OF BIRTH (Mo/Day/Yr)
	Michelle Olsen	04/09/1995
	8c. MOTHER'S NAME PRIOR TO FIRST MARRIAGE (First, Middle, Last, Suffix)	8d. BIRTHPLACE (State, Territory, or Foreign Country)
	Michelle Olsen	New Hampshire

9a. RESIDENCE OF MOTHER-STATE	9b. COUNTY	9c. CITY, TOWN, OR LOCATION
Florida	Hillsborough	Tampa

9d. STREET AND NUMBER	9e. APT. NO.	9f. ZIP CODE	9g. INSIDE CITY LIMITS?
1045 Ponderosa Drive	-	33647	■ Yes □ No

YES □ NO □

U.S. STANDARD CERTIFICATE OF LIVE BIRTH

LOCAL FILE NO. | BIRTH NUMBER:

CHILD	1. CHILD'S NAME (First, Middle, Last, Suffix)	2. TIME OF BIRTH	3. SEX	4. DATE OF BIRTH (Mo/Day/Yr)
	Gregory Baker	1:45	M	05/06/2011

5. FACILITY NAME (If not institution, give street and number)	6. CITY, TOWN, OR LOCATION OF BIRTH	7. COUNTY OF BIRTH
Tampa General Hospital	Tampa	Hillsborough

MOTHER	8a. MOTHER'S CURRENT LEGAL NAME (First, Middle, Last, Suffix)	8b. DATE OF BIRTH (Mo/Day/Yr)
	Allison Baker	03/25/1993
	8c. MOTHER'S NAME PRIOR TO FIRST MARRIAGE (First, Middle, Last, Suffix)	8d. BIRTHPLACE (State, Territory, or Foreign Country)
	Allison Ball	Alabama

9a. RESIDENCE OF MOTHER-STATE	9b. COUNTY	9c. CITY, TOWN, OR LOCATION
Florida	Hillsborough	Tampa

9d. STREET AND NUMBER	9e. APT. NO.	9f. ZIP CODE	9g. INSIDE CITY LIMITS?
584 Capwood Avenue	-	33618	■ Yes □ No

YES □ NO □

U.S. STANDARD CERTIFICATE OF LIVE BIRTH

LOCAL FILE NO. BIRTH NUMBER:

CHILD

1. CHILD'S NAME (First, Middle, Last, Suffix)	2. TIME OF BIRTH	3. SEX	4. DATE OF BIRTH (Mo/Day/Yr)
Aubrey Burton	1:19	F	05/09/2011

5. FACILITY NAME (If not institution, give street and number)	6. CITY, TOWN, OR LOCATION OF BIRTH	7. COUNTY OF BIRTH
Tampa General Hospital	Tampa	Hillsborough

MOTHER

8a. MOTHER'S CURRENT LEGAL NAME (First, Middle, Last, Suffix)	8b. DATE OF BIRTH (Mo/Day/Yr)
Caitlin Marsh	12/18/1996
8c. MOTHER'S NAME PRIOR TO FIRST MARRIAGE (First, Middle, Last, Suffix)	**8d. BIRTHPLACE (State, Territory, or Foreign Country)**
Caitlin Marsh	Washington

9a. RESIDENCE OF MOTHER-STATE	9b. COUNTY	9c. CITY, TOWN, OR LOCATION
Florida	Hillsborough	Tampa

9d. STREET AND NUMBER	9e. APT. NO.	9f. ZIP CODE	9g. INSIDE CITY LIMITS?
889 Country Road	-	33685	■ Yes □ No

YES ☐ NO ☐

U.S. STANDARD CERTIFICATE OF LIVE BIRTH

LOCAL FILE NO. BIRTH NUMBER:

CHILD

1. CHILD'S NAME (First, Middle, Last, Suffix)	2. TIME OF BIRTH	3. SEX	4. DATE OF BIRTH (Mo/Day/Yr)
Jason McGraw	04:45	M	05/19/2011

5. FACILITY NAME (If not institution, give street and number)	6. CITY, TOWN, OR LOCATION OF BIRTH	7. COUNTY OF BIRTH
Florida Hospital	Orlando	Orange

MOTHER

8a. MOTHER'S CURRENT LEGAL NAME (First, Middle, Last, Suffix)	8b. DATE OF BIRTH (Mo/Day/Yr)
Sabrina McGraw	02/06/1994
8c. MOTHER'S NAME PRIOR TO FIRST MARRIAGE (First, Middle, Last, Suffix)	**8d. BIRTHPLACE (State, Territory, or Foreign Country)**
Sabrina Harding	Nebraska

9a. RESIDENCE OF MOTHER-STATE	9b. COUNTY	9c. CITY, TOWN, OR LOCATION
Florida	Orange	Orlando

9d. STREET AND NUMBER	9e. APT. NO.	9f. ZIP CODE	9g. INSIDE CITY LIMITS?
5652 Alafaya Drive	-	32810	■ Yes □ No

YES ☐ NO ☐

U.S. STANDARD CERTIFICATE OF LIVE BIRTH

LOCAL FILE NO. BIRTH NUMBER:

CHILD

1. CHILD'S NAME (First, Middle, Last, Suffix)	2. TIME OF BIRTH	3. SEX	4. DATE OF BIRTH (Mo/Day/Yr)
Savannah Marie Price	10:39	F	06/26/2011

5. FACILITY NAME (If not institution, give street and number)	6. CITY, TOWN, OR LOCATION OF BIRTH	7. COUNTY OF BIRTH
Tampa General Hospital	Tampa	Hillsborough

MOTHER

8a. MOTHER'S CURRENT LEGAL NAME (First, Middle, Last, Suffix)	8b. DATE OF BIRTH (Mo/Day/Yr)
Julie Price	01/30/1997
8c. MOTHER'S NAME PRIOR TO FIRST MARRIAGE (First, Middle, Last, Suffix)	**8d. BIRTHPLACE (State, Territory, or Foreign Country)**
Julie Mills	Florida

9a. RESIDENCE OF MOTHER-STATE	9b. COUNTY	9c. CITY, TOWN, OR LOCATION
Florida	Hillsborough	Tampa

9d. STREET AND NUMBER	9e. APT. NO.	9f. ZIP CODE	9g. INSIDE CITY LIMITS?
344 Spring Tree Drive	-	33624	■ Yes □ No

YES ☐ NO ☐

U.S. STANDARD CERTIFICATE OF LIVE BIRTH

LOCAL FILE NO. BIRTH NUMBER:

CHILD	1. CHILD'S NAME (First, Middle, Last, Suffix) Kyle Palmer	2. TIME OF BIRTH 17:25	3. SEX M	4. DATE OF BIRTH (Mo/Day/Yr) 04/08/2011
	5. FACILITY NAME (If not institution, give street and number) Tampa General Hospital	6. CITY, TOWN, OR LOCATION OF BIRTH Tampa	7. COUNTY OF BIRTH Hillsborough	
MOTHER	8a. MOTHER'S CURRENT LEGAL NAME (First, Middle, Last, Suffix) Lauren Palmer	8b. DATE OF BIRTH (Mo/Day/Yr) 02/30/1990		
	8c. MOTHER'S NAME PRIOR TO FIRST MARRIAGE (First, Middle, Last, Suffix) Lauren Sims	8d. BIRTHPLACE (State, Territory, or Foreign Country) Florida		
	9a. RESIDENCE OF MOTHER-STATE Florida	9b. COUNTY Hillsborough	9c. CITY, TOWN, OR LOCATION Tampa	
	9d. STREET AND NUMBER 215 Allen Lane	9e. APT. NO.	9f. ZIP CODE 33660	9g. INSIDE CITY LIMITS? ■ Yes □ No

YES ☐ NO ☐

U.S. STANDARD CERTIFICATE OF LIVE BIRTH

LOCAL FILE NO. BIRTH NUMBER:

CHILD	1. CHILD'S NAME (First, Middle, Last, Suffix) Brian Barnes	2. TIME OF BIRTH 13:21	3. SEX M	4. DATE OF BIRTH (Mo/Day/Yr) 12/30/2011
	5. FACILITY NAME (If not institution, give street and number) Tampa General Hospital	6. CITY, TOWN, OR LOCATION OF BIRTH Tampa	7. COUNTY OF BIRTH Hillsborough	
MOTHER	8a. MOTHER'S CURRENT LEGAL NAME (First, Middle, Last, Suffix) Alyssa Adams	8b. DATE OF BIRTH (Mo/Day/Yr) 02/17/1994		
	8c. MOTHER'S NAME PRIOR TO FIRST MARRIAGE (First, Middle, Last, Suffix) Alyssa Adams	8d. BIRTHPLACE (State, Territory, or Foreign Country) Florida		
	9a. RESIDENCE OF MOTHER-STATE Florida	9b. COUNTY Hillsborough	9c. CITY, TOWN, OR LOCATION Tampa	
	9d. STREET AND NUMBER 3352 Navajo Avenue	9e. APT. NO. -	9f. ZIP CODE 33688	9g. INSIDE CITY LIMITS? ■ Yes □ No

YES ☐ NO ☐

U.S. STANDARD CERTIFICATE OF LIVE BIRTH

LOCAL FILE NO. BIRTH NUMBER:

CHILD	1. CHILD'S NAME (First, Middle, Last, Suffix) Julia Roberts	2. TIME OF BIRTH 5:21	3. SEX F	4. DATE OF BIRTH (Mo/Day/Yr) 12/12/2011
	5. FACILITY NAME (If not institution, give street and number) Tampa General Hospital	6. CITY, TOWN, OR LOCATION OF BIRTH Tampa	7. COUNTY OF BIRTH Hillsborough	
MOTHER	8a. MOTHER'S CURRENT LEGAL NAME (First, Middle, Last, Suffix) Kelly Carson	8b. DATE OF BIRTH (Mo/Day/Yr) 06/25/1995		
	8c. MOTHER'S NAME PRIOR TO FIRST MARRIAGE (First, Middle, Last, Suffix) Kelly Carson	8d. BIRTHPLACE (State, Territory, or Foreign Country) Florida		
	9a. RESIDENCE OF MOTHER-STATE Florida	9b. COUNTY Hillsborough	9c. CITY, TOWN, OR LOCATION Tampa	
	9d. STREET AND NUMBER 3315 Morris Bridge Road	9e. APT. NO. -	9f. ZIP CODE 33631	9g. INSIDE CITY LIMITS? ■ Yes □ No

YES ☐ NO ☐

U.S. STANDARD CERTIFICATE OF LIVE BIRTH

LOCAL FILE NO. | BIRTH NUMBER:

CHILD	1. CHILD'S NAME (First, Middle, Last, Suffix) Natalie Portman	2. TIME OF BIRTH 00:25	3. SEX F	4. DATE OF BIRTH (Mo/Day/Yr) 01/01/2012
	5. FACILITY NAME (If not institution, give street and number) Tampa General Hospital	6. CITY, TOWN, OR LOCATION OF BIRTH Tampa	7. COUNTY OF BIRTH Hillsborough	
MOTHER	8a. MOTHER'S CURRENT LEGAL NAME (First, Middle, Last, Suffix) Amber Jane Portman	8b. DATE OF BIRTH (Mo/Day/Yr) 05/23/1995		
	8c. MOTHER'S NAME PRIOR TO FIRST MARRIAGE (First, Middle, Last, Suffix) Amber Jane Moore	8d. BIRTHPLACE (State, Territory, or Foreign Country) Florida		
	9a. RESIDENCE OF MOTHER-STATE Florida	9b. COUNTY Hillsborough	9c. CITY, TOWN, OR LOCATION Tampa	
	9d. STREET AND NUMBER 1151 Davis Road	9e. APT. NO. -	9f. ZIP CODE 33615	9g. INSIDE CITY LIMITS? ■ Yes □ No

YES ☐ NO ☐

U.S. STANDARD CERTIFICATE OF LIVE BIRTH

LOCAL FILE NO. | BIRTH NUMBER:

CHILD	1. CHILD'S NAME (First, Middle, Last, Suffix) Thomas Edison	2. TIME OF BIRTH 00:02	3. SEX M	4. DATE OF BIRTH (Mo/Day/Yr) 01/01/2011
	5. FACILITY NAME (If not institution, give street and number) Tampa General Hospital	6. CITY, TOWN, OR LOCATION OF BIRTH Tampa	7. COUNTY OF BIRTH Hillsborough	
MOTHER	8a. MOTHER'S CURRENT LEGAL NAME (First, Middle, Last, Suffix) Cassandra Edison	8b. DATE OF BIRTH (Mo/Day/Yr) 12/01/1994		
	8c. MOTHER'S NAME PRIOR TO FIRST MARRIAGE (First, Middle, Last, Suffix) Cassandra Franklin	8d. BIRTHPLACE (State, Territory, or Foreign Country) Florida		
	9a. RESIDENCE OF MOTHER-STATE Florida	9b. COUNTY Hillsborough	9c. CITY, TOWN, OR LOCATION Tampa	
	9d. STREET AND NUMBER 593 Hibiscus Drive	9e. APT. NO. -	9f. ZIP CODE 33619	9g. INSIDE CITY LIMITS? ■ Yes □ No

YES ☐ NO ☐

U.S. STANDARD CERTIFICATE OF LIVE BIRTH

LOCAL FILE NO. | BIRTH NUMBER:

CHILD	1. CHILD'S NAME (First, Middle, Last, Suffix) Angelina Jolie	2. TIME OF BIRTH 21:19	3. SEX F	4. DATE OF BIRTH (Mo/Day/Yr) 10/30/2011
	5. FACILITY NAME (If not institution, give street and number) Brandon Regional Hospital	6. CITY, TOWN, OR LOCATION OF BIRTH Brandon	7. COUNTY OF BIRTH Hillsborough	
MOTHER	8a. MOTHER'S CURRENT LEGAL NAME (First, Middle, Last, Suffix) Kera Jolie	8b. DATE OF BIRTH (Mo/Day/Yr) 11/31/1993		
	8c. MOTHER'S NAME PRIOR TO FIRST MARRIAGE (First, Middle, Last, Suffix) Kera Jolie	8d. BIRTHPLACE (State, Territory, or Foreign Country) Oklahoma		
	9a. RESIDENCE OF MOTHER-STATE Florida	9b. COUNTY Hillsborough	9c. CITY, TOWN, OR LOCATION Tampa	
	9d. STREET AND NUMBER 898 Walker Road	9e. APT. NO. 101	9f. ZIP CODE 33612	9g. INSIDE CITY LIMITS? ■ Yes □ No

YES ☐ NO ☐

Birth Rates in Florida Assignment

Epidemiology in real life

As an epidemiologist, you may need data that are not already put together for you. You have two main options: collect the data yourself by conducting a research study or relaying on data that have already been collected for a number of purposes. If you use pre-existing data, you will likely need to find things from a variety of sources, including health department statistics, individual study data, the U.S. Census Bureau, the CDC, etc. You need to be able to sift through extensive data sets and pick out the information that allows you to answer a research question.

Many epidemiologists work in state health departments and as such, they are often asked to provide information fairly rapidly. In that situation most of the time they use data that is already available, and the greatest source of information is on the WEB. This assignment will give you an opportunity to understand how you, as a new epidemiologist, might use the internet to help answer important research questions.

New job, day 1

Today is June 7th and you graduated with a Masters in Public Health from the University of South Florida on May 26th. You are really excited as you got a position in the Maternal Child Health Bureau at the Health Department but you are also pretty nervous about this new job. Following a weeklong celebration in Cancun, you are ready for your first full day. You spent nearly all day yesterday filling out health insurance forms, getting a computer password, finding your office, getting lost looking for a parking space, and meeting your supervisor, Joan Jetson. Joan seems really nice but she is also quite noticeably pregnant. The plan for today is for her to review with you the types of data they have and to explain the projects you will be assigned to. You are feeling pretty comfortable as Joan has a plan for training you but she also told you she was short staffed due to vacations so you might need to get involved fairly quickly. The atmosphere felt good, people were friendly and relaxed, and you are fairly confident you can do a good job.

You arrived about 10 minutes late because you got lost trying to park and so you ran up to your department. But no one was there. It felt like one of those dreams you had when you went to take a final and you were in the wrong place. Just as you were getting really nervous, people starting coming in from a meeting but everyone looked somber. Paul, the health education outreach director, threw down the morning newspaper, and said, "Just look at this!!!" The headline jumped out at you.

Hillsborough County has the highest teenage pregnancy in the nation. Is our health department failing us???

Paul said the mayor is furious and we are going to have to get the data ready for a meeting with him at 1pm this afternoon to prepare him for a response to the media. "With all the recent budget cuts, this is the last thing we need! Then on top of everything else, Joan came in at 7am to work on this and went into labor. She is now in the maternity ward at Tampa Community Hospital and we don't have anyone who can help out. I have a meeting in Plant City with the WIC staff so you are just going to have to prepare this report and meet with the mayor yourself. There is no one else who can do it. Joan left some notes by her desk and the mayor wants to see you at 1pm sharp. You do have your computer password, don't you?" he said as he was running out the door.

You walked over to the computer and found a note from Joan. "I had to run but here is the memo from the health director. Just pull together all the information the mayor might possibly need and be prepared to answer any questions he might have. He will meet with you at 1:00 pm and he has a press conference at 1:40pm so you won't have much time to go back and get more information. I also left you a few notes that might help out, and I started a table that will give you the information I need. Good luck!"

You will have 30 minutes to brief the mayor. Be sure you have all your information with you when you go into that meeting. The mayor's questions (the briefing) will be located in Blackboard. You can use your notes and the computer to find additional information but you will only have 30 minutes from the start to complete all the responses. Make sure that you are comfortable with your data before you go in and answer the questions. Remember, the mayor's time is limited! The following documents have been provided to help you answer any questions but don't hesitate to use other sources to get more information.

Documents Provided:

1) Memorandum from the director of the health department to Joan Jetson
2) 2) Joan's notes
3) Table 1. Data Set: Florida Department of Health, Office of Vital Statistics Chart – "Resident Births: Frequencies
4) Data Set: Florida Department of Health, Office of Vital Statistics Chart- " Table B-12 : Residents Live Births To Mothers 15 Through 19 Years Of Age For Selected Indicators, By County, Florida
5) Worksheet with example

*(*Please note: These are real databases and not always alphabetized in the same order.)*

Formulas:

Crude Birth Rate

(Number of live births during a specified time period / mid-interval population of that year) x 1000

This mid-interval population includes males and females.

Age Specific Birth Rate

(Number of births for an age group in a county during a specified time period/ population of women in that age group in the county that year) x 1000

Important comment

Please note the difference in the denominator between the two measures. The crude birth rate includes both genders and people of all ages, while the age specific birth rate is limited to women of that age group. While this may seems surprising, it actually makes quite a bit of sense. The crude birth rate is usually used to evaluate population growth when comparing countries. It is a demographic measure of changes in a country. It would be very difficult to limit it to just women of childbearing age so the larger population is used.

The age-specific birth rate is usually used for program planning and identifying risk factors for pregnancy. It thus makes more sense to focus on the population in which the births would occur.

Since there is a difference in the denominator used, there would be a large difference between the crude birth rate and the age specific birth rate. The crude birth rate will be smaller because the denominator includes men and people either too old or too young to get pregnant, while the age-specific denominator only includes people "at risk" of becoming pregnant.

Memorandum

TO: Joan Jetson, Director the maternal health Bureau

FROM: Will Robinson, Director of the Hillsborough County Health Department

SUBJECT: Newspaper article

IMPORTANCE: High

The mayor is extremely concerned about the Newspaper article. You need to meet with the mayor at 1:00 pm today so you can provide him with the information he needs to respond to the press. You will only have 30 minutes to respond so try to anticipate any questions he might have. Please bring anything you have that can provide the mayor with the information he will need to respond to the press. He has a press conference scheduled for 1:40. I know I don't need to tell you how important this is, as the mayor is reviewing our budget next week. I am already committed to presenting our budget at a meeting in Tallahassee at 3pm but I am sure you can handle it.

I suggest you have the following information with you:

Crude Birth Rates in Florida for **all** Florida counties.

Age Specific Rates for teenage births for **all Florida Counties.**

Rates ranked by size so you can easily provide those with the highest and lowest rates

Numbers ranked by size so you can identify counties with the highest and lowest numbers of adolescent births

The overall Florida birth rate, as well as the Florida teenage birth rate

The definition of a teenage birth. You should be able to get that at the National Center for Health Statistics

Anything else you think is important.

You might bring your new assistant along as this could be an interesting learning experience.

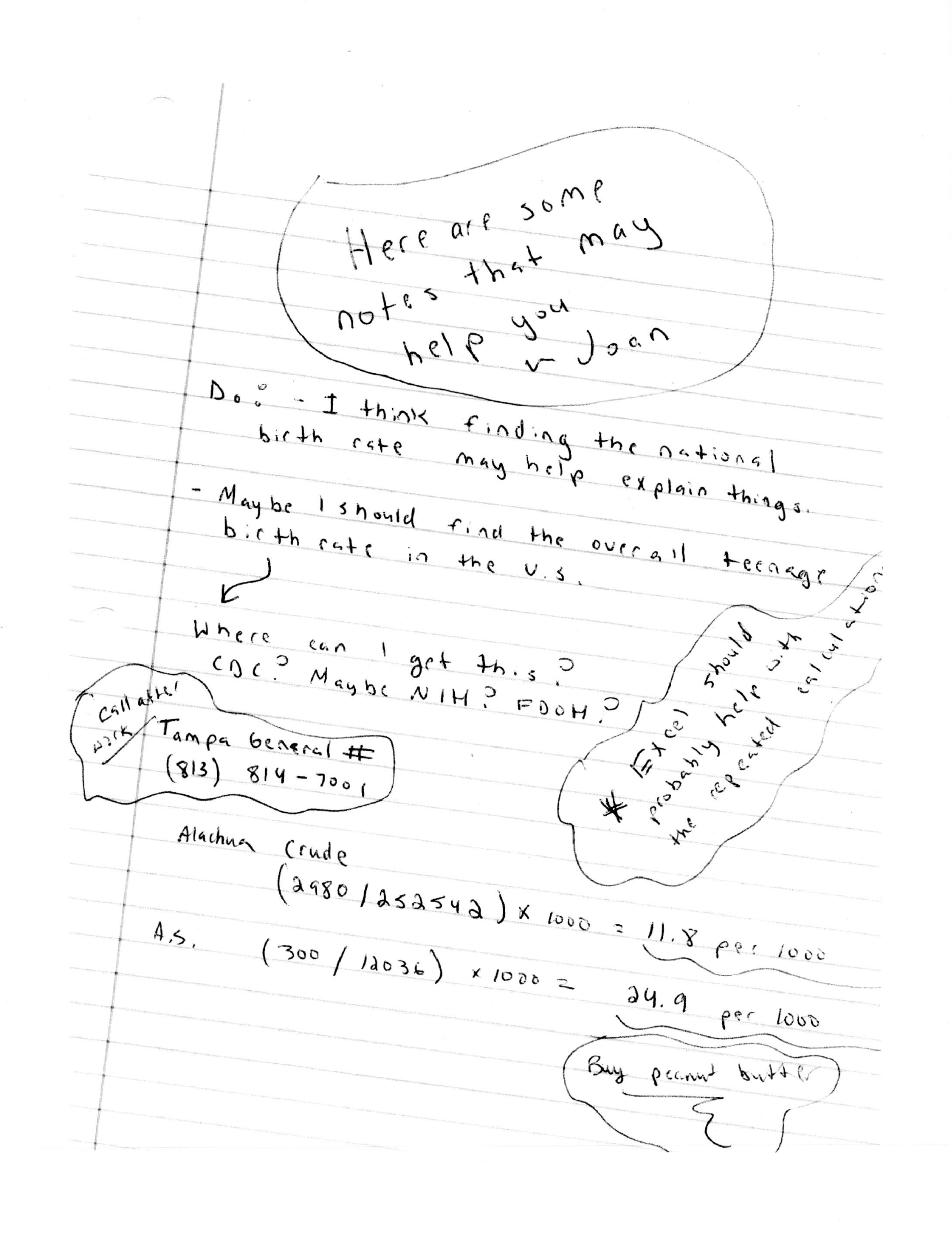
Here are some notes that may help you
~ Joan
Do: - I think finding the national birth rate may help explain things.
- Maybe I should find the overall teenage birth rate in the U.S.
Where can I get this? CDC? Maybe NIH? FDOH?
Call after work
Tampa General #
(813) 814-7001
* Excel should probably help with the repeated calculation.
Alachua Crude
(2980 / 252542) x 1000 = 11.8 per 1000
A.S. (300 / 12036) x 1000 = 24.9 per 1000
Buy peanut butter

Table 1. Florida Department of Health, Office of Vital Statistics – "Resident Births: Frequencies (2014)

	Resident Births								
	0-14	15-19	20-24	25-29	30-34	35-39	40-44	45+	Total
Alachua	0	147	600	916	840	344	64	0	2,916
Baker	0	50	121	105	60	25	4	0	365
Bay	1	179	709	717	496	182	42	0	2,328
Bradford	1	31	99	78	57	12	4	0	282
Brevard	3	283	1,202	1,674	1,337	610	140	0	5,259
Broward	7	906	3,852	6,103	6,653	3,710	914	1	22,213
Calhoun	0	22	38	41	20	6	4	0	131
Charlotte	0	75	300	299	220	86	26	0	1,007
Citrus	0	89	364	296	178	72	17	0	1,016
Clay	0	135	531	666	474	226	48	0	2,083
Collier	0	189	634	926	922	477	127	0	3,288
Columbia	0	96	285	240	136	65	9	0	832
Miami-Dade	13	1,318	6,027	8,912	9,076	5,216	1,324	2	31,990
Desoto	1	49	121	121	62	27	3	0	384
Dixie	0	13	56	57	36	4	3	0	169
Duval	13	778	3,141	3,893	3,132	1,272	256	1	12,514
Escambia	5	312	1,154	1,186	826	342	50	0	3,880
Flagler	0	56	199	244	206	98	25	0	833
Franklin	0	14	43	31	10	2	0	0	100
Gadsden	1	45	172	148	99	58	12	0	535
Gilchrist	0	18	61	49	29	10	0	0	167
Glades	0	10	18	17	9	4	2	0	60
Gulf	0	11	46	29	17	10	3	0	117
Hamilton	1	13	55	44	17	13	2	0	145
Hardee	1	60	138	116	67	26	6	0	414
Hendry	1	56	171	177	106	45	10	0	569
Hernando	0	91	429	456	332	145	33	0	1,488
Highlands	1	90	279	302	170	73	21	0	937
Hillsborough	28	1,057	3,930	4,788	4,510	2,048	443	1	16,846
Holmes	1	29	64	51	43	18	1	0	207
Indian River	0	81	340	370	300	145	45	0	1,282
Jackson	0	51	185	149	91	32	6	0	514

	Resident Births								
	0-14	**15-19**	**20-24**	**25-29**	**30-34**	**35-39**	**40-44**	**45+**	**Total**
Jefferson	0	4	38	45	22	14	4	0	127
Lafayette	0	6	26	24	15	5	0	0	76
Lake	2	238	868	939	722	329	64	1	3,166
Lee	4	435	1,492	1,859	1,610	771	173	1	6,352
Leon	5	185	716	877	843	386	67	0	3,085
Levy	1	33	140	113	92	23	7	0	409
Liberty	0	7	37	21	17	10	0	0	92
Madison	0	10	60	65	35	19	2	0	191
Manatee	8	300	881	978	905	386	83	0	3,545
Marion	6	272	1,023	1,053	691	302	64	0	3,417
Martin	2	79	246	376	346	170	43	0	1,263
Monroe	1	31	147	230	200	115	24	0	749
Nassau	0	46	201	244	174	60	19	0	745
Okaloosa	2	147	715	974	662	281	41	0	2,827
Okeechobee	1	57	184	181	85	40	5	0	553
Orange	11	884	3,410	4,577	4,517	2,268	525	0	16,221
Osceola	5	325	1,020	1,259	971	482	121	0	4,195
Palm Beach	14	660	2,674	3,883	4,174	2,403	585	0	14,433
Pasco	3	290	1,149	1,417	1,237	593	132	0	4,826
Pinellas	8	468	1,951	2,546	2,272	1,016	237	0	8,519
Polk	7	647	2,191	2,322	1,635	682	116	0	7,608
Putnam	1	96	283	241	131	63	15	0	831
Saint Johns	0	70	302	640	733	331	65	0	2,148
Saint Lucie	1	169	731	896	759	311	98	0	2,969
Santa Rosa	1	106	420	609	468	180	35	0	1,822
Sarasota	2	171	609	876	812	395	86	1	2,955
Seminole	4	173	835	1,336	1,342	666	150	0	4,515
Sumter	0	48	184	133	86	34	8	0	494
Suwannee	1	48	142	141	72	24	12	0	440
Taylor	0	22	82	68	32	11	2	0	217
Union	0	12	49	52	28	12	1	0	154
Volusia	1	313	1,270	1,470	1,111	474	125	0	4,767
Wakulla	0	26	87	115	76	27	4	0	335
Walton	0	47	176	244	171	69	18	0	725
Washington	0	30	96	61	37	15	4	0	243
Unknown	0	2	4	4	6	3	1	0	20
Total	**169**	**12,811**	**49,803**	**64,070**	**57,620**	**28,373**	**6,580**	**8**	**219,905**

Table B-12. Resident population in Florida counties by race and age

	White		Black				
	Female	Male	Female	Male	Unknown	Total	Females 15-19
Alachua	21,652	23,026	13,073	13,541	32	247,336	12,647
Baker	3,737	3,974	639	668	7	27,115	871
Bay	21,993	23,084	5,467	5,666	708	168,852	5,285
Bradford	3,184	3,530	918	913	0	28,520	726
Brevard	54,716	58,250	12,137	12,345	137	543,376	16,299
Broward	165,824	175,662	105,821	109,089	2,495	1,748,066	55,278
Calhoun	1,670	1,707	311	290	0	14,625	427
Charlotte	12,130	12,714	1,228	1,354	15	159,978	3,529
Citrus	11,414	12,150	822	865	13	141,236	3,428
Clay	21,399	22,579	3,540	3,746	65	190,865	7,252
Collier	34,985	36,625	5,566	5,742	21	321,520	8,395
Columbia	7,752	8,064	2,309	2,316	7	67,531	2,045
Miami-Dade	295,134	311,014	123,771	128,435	2,297	2,496,435	82,420
Desoto	4,591	4,845	791	853	2	34,862	981
Dixie	1,907	1,978	208	197	1	16,422	411
Duval	98,047	103,299	64,123	66,962	441	864,263	28,707
Escambia	34,087	36,058	18,775	19,045	36	297,619	10,724
Flagler	5,904	6,289	1,244	1,348	27	95,696	2,704
Franklin	1,215	1,340	208	223	14	11,549	249
Gadsden	3,340	3,424	5,767	5,931	14	46,389	1,582
Gilchrist	2,088	2,172	139	138	3	16,939	520
Glades	802	823	246	299	0	12,884	290
Gulf	1,439	1,517	305	323	19	15,863	349
Hamilton	1,304	1,371	845	853	0	14,799	393
Hardee	5,418	5,695	420	440	5	27,731	922
Hendry	6,737	7,096	1,457	1,579	15	39,140	1,516
Hernando	14,852	15,367	1,655	1,664	49	172,778	4,897
Highlands	9,321	9,676	2,438	2,640	12	98,786	2,470
Hillsborough	149,370	156,730	49,811	51,953	805	1,229,226	43,319
Holmes	2,524	2,706	141	149	6	19,927	588
Indian River	11,592	12,130	2,916	2,971	1,137	138,028	3,629
Jackson	4,996	5,115	2,049	2,167	8	49,746	1,520
Jefferson	990	1,102	1,001	1,045	2	14,761	378

	White		Black				
	Female	Male	Female	Male	Unknown	Total	Females 15-19
Lafayette	963	930	118	113	0	8,870	220
Lake	27,840	29,393	6,109	6,287	47	297,052	8,392
Lee	60,167	63,704	11,482	11,809	109	618,754	17,174
Leon	21,808	23,111	16,766	17,282	87	275,487	14,722
Levy	4,334	4,542	837	880	1	40,801	1,256
Liberty	938	1,032	123	135	0	8,365	203
Madison	1,468	1,532	1,478	1,594	1	19,224	609
Manatee	34,850	36,030	7,692	8,030	60	322,833	8,861
Marion	30,761	32,245	8,894	9,321	20	331,298	9,214
Martin	12,739	13,142	3,085	3,382	40	146,318	3,709
Monroe	9,045	9,573	1,286	1,365	49	73,090	1,497
Nassau	8,625	9,006	948	985	15	73,314	2,268
Okaloosa	27,241	28,087	5,783	6,300	54	180,822	5,540
Okeechobee	6,254	6,613	750	741	31	39,996	1,303
Orange	125,433	132,333	57,607	59,961	243	1,145,956	43,512
Osceola	32,617	34,476	5,395	5,654	44	268,685	10,300
Palm Beach	122,967	130,349	54,296	56,234	357	1,320,134	38,595
Pasco	48,287	50,764	4,171	4,481	209	464,697	13,401
Pinellas	92,728	98,049	26,208	27,375	569	916,542	24,574
Polk	70,245	74,011	20,236	21,150	78	602,095	19,666
Putnam	8,635	9,504	3,283	3,366	21	74,364	2,299
Saint Johns	16,617	17,399	2,730	2,721	55	190,039	6,214
Saint Lucie	23,606	25,251	10,337	10,914	111	277,789	8,527
Santa Rosa	18,850	19,822	1,744	1,875	17	151,372	5,025
Sarasota	31,701	33,421	4,505	4,913	65	379,448	8,840
Seminole	46,784	49,477	11,148	11,521	83	422,718	15,081
Sumter	4,669	4,829	1,299	1,347	6	93,420	1,219
Suwannee	4,727	4,950	1,016	1,004	6	41,551	1,216
Taylor	2,420	2,608	839	766	3	22,570	623
Union	1,692	1,694	300	312	1	15,535	390
Volusia	50,059	52,391	11,141	11,216	112	494,593	14,782
Wakulla	3,001	3,216	482	511	2	30,776	940
Walton	6,038	6,511	656	677	16	55,043	1,438
Washington	2,495	2,614	610	659	12	24,896	734
Total	1,976,919	2,083,923	713,567	740,715	10,958	18,801,310	597,095

Table 3. Worksheet for completing the assignment.

County	All Mothers giving birth	Total Population	Total Births (15-19 years)	Total female population (15-19 years)	Overall birth rate	Age specific birth rate (15-19 years)
Alachua	2,916	247,336	147	12,647	11.8	11.6
Baker	365	27,115	50	871	13.5	57.4
Bay						
Bradford						
Brevard						
Broward						
Calhoun						
Charlotte						
Citrus						
Clay						
Collier						
Columbia						
Miami-Dade						
Desoto						
Dixie						
Duval						
Escambia						
Flagler						
Franklin						
Gadsden						
Gilchrist						
Glades						
Gulf						
Hamilton						
Hardee						
Hendry						
Hernando						

Highlands						
Hillsborough						
Holmes						
Indian River						
Jackson						
Jefferson						
Lafayette						
Lake						
Lee						
Leon						
Levy						
Liberty						
Madison						
Manatee						
Marion						
Martin						
Monroe						
Nassau						
Okaloosa						
Okeechobee						
Orange						
Osceola						
Palm Beach						
Pasco						
Pinellas						
Polk						
Putnam						
Saint Johns						
Saint Lucie						
Santa Rosa						
Sarasota						
Seminole						

Sumter						
Suwannee						
Taylor						
Union						
Volusia						
Wakulla						
Walton						
Washington						
Unknown						

Study Designs

We have reviewed a number of different study designs. It can be challenging to identify the different types of study designs. The exercise below provides you with 5 research studies, all addressing the association of intake of dietary iron and anemia. Each question is worded in such a way that it dictates a specific form of analysis and means of gathering data (e.g. study design). For each of the specific research question, decide which study design MOST APPROPRIATELY answers the question? Indicate your choice on the answer sheet by writing in ONE of the following study designs.

Cross Sectional

Ecologic

Cohort

Case Control

Clinical trial

1. Are individuals with a low intake of dietary iron more likely to develop anemia than individuals with a high intake of dietary iron?

2. Among premenopausal women, is their dietary iron intake associated with their hemoglobin level?

3. Is the mean per capita intake of dietary iron associated with the prevalence of anemia in a population?

4. Are teenage girls with anemia more likely to have had a history of a low intake of dietary iron than teenage girls without anemia?

5. Among women with anemia, will intake of a dietary iron supplement result in normal hemoglobin levels?

Module 6: Causation and Experimental Studies

The following exercises will give you experience in determining causality in epidemiological research and in developing research studies. These exercises are simulations of the type of experiences epidemiologists may have when addressing research problems in the community. Take the time to work out these problems as actually doing the exercises will help improve your epidemiological skills. Good luck and have fun.

Causality exercise

Determining causality is an important function in epidemiology as one of the primary goals of epidemiology is to identify the causes of diseases so that prevention and treatment programs can be designed and implemented.

There are several models of causality. The first model was presented by Robert Koch in 1882. These are presented below:

1. Koch's Postulates

1. The organism must be observed in every case of disease.
2. The organism must be isolated and grown in pure culture.
3. The pure culture must, when inoculated into a susceptible animal, reproduce the disease.
4. The organism, must be observed in, and recovered from, the experimental animal1.

Clearly these postulates apply best to infectious disease and they were very important at the time researchers were trying to identify the cause and potential treatment of infectious diseases. It would be much more difficult to apply these postulates to chronic diseases such as the association of tobacco smoke and lung disease or heart disease.

The next major step in identifying causality was based on the work of an expert committee appointed by the U.S Surgeon General to review the evidence of an association between smoking and cancer. These guidelines for assessing causality have been revised a number of times and are summarized briefly below2. Many times different epidemiologists use different terms to describe guidelines for causality. Some are identified in the list.

2. Guidelines from the Expert Committee on Causality

1. Temporal relationship: The exposure must occur prior to the disease.
2. Strength of the association: The stronger an association, the more likely it is causal. This refers to a larger odds ratio or relative risk.
3. Dose-response (also known as Biological gradient): As the dose increases, the risk of disease increases.
4. Replication of findings: If an association is causal, it will be found in multiple studies and in different populations.
5. Biologic plausibility (also known as Coherence): The association needs to be coherent with our current biological knowledge.
6. Consideration of alternative explanations: Ruling out other possibilities increasing our confidence an association is causal.
7. Cessation of exposure: The risk of disease declines when the exposure is removed.
8. Specificity of exposure: A specific exposure is associated with only one disease. This guideline is generally applicable for infectious disease but is not appropriate in chronic disease where one exposure such as smoking may cause a number of diseases such as cancer, heart disease, and stroke.

9. Consistency with other knowledge (also known as Analogy): This refers to a finding being consistent with other data, including other studies, and changes in disease that show overall consistency.

Other discussions of causality consider other models: the epidemiologic triangle and the web of causation. The epidemiologic triangle is concerned with the relationship between an agent, the host, and environment (Figure 1). Again this model is effective for infectious diseases but does not apply as well to chronic disease, where there are often multiple risk factors.

3. The epidemiologic triangle

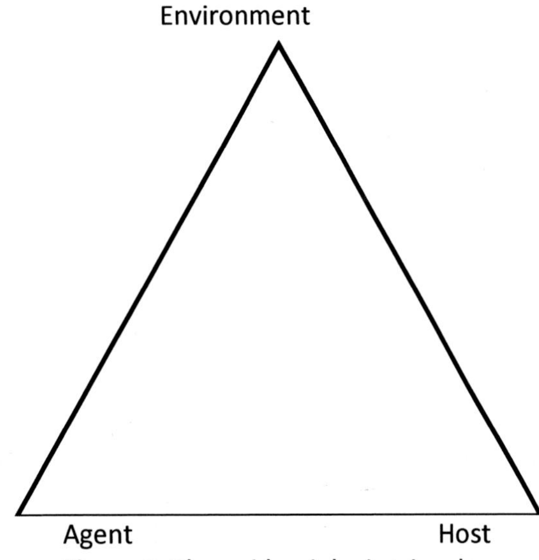

Figure 1. The epidemiologic triangle

4. The WEB of causation

The web of causation is used to describe the association of a number of risk factors with a disease. This model may be used for an infectious disease as it takes into account various factors other than an infectious agent that may cause a disease. This may include genetic susceptibility, nutritional status, and other illnesses. However, this model is often used in describing a chronic disease in which a number of factors may result in an illness (Figure 2).

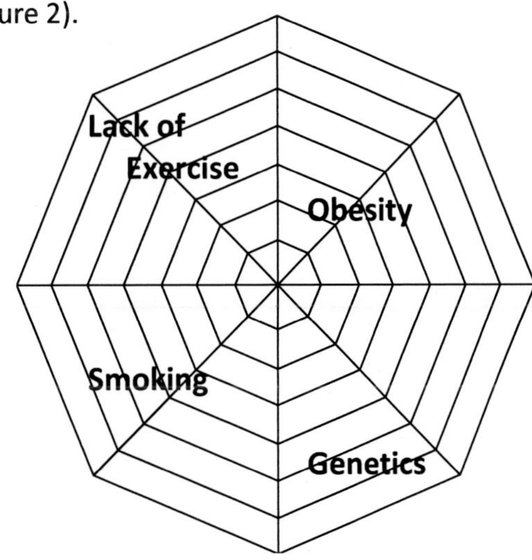

Figure 2. The Web of Causation for a heart disease

5. Model of Necessary and Sufficient Causes of Disease

Another model presented is that of necessary and sufficient causes. According to this model, there are the following 4 types of causal relationships:

1. Necessary and sufficient
    a. A factor always causes a disease
    b. The disease cannot occur without that factor
        i. This situation does not occur in reality
2. Necessary but not sufficient
    a. Multiple factors are required to cause a disease
    b. Each factor is necessary but is not sufficient in itself to cause a disease
        i. TB requires a TB bacillus but also needs a susceptible host
3. Sufficient but not necessary
    a. A factor can cause disease but other factors can also do so in the absence of that factor
    i. Smoking or obesity can lead to heart disease but each can do in the absence of the other
4. Neither sufficient nor necessary
    a. Any factor by itself is not necessary or sufficient
    b. Multiple factors cause the disease
        i. This is the situation that occurs in most chronic diseases

The following exercise will give you an opportunity to use these concepts of causation in an applied setting.

A search for the cause of the mystery illness

It is June of the year 2025, and the world's population has been decimated by the TWD zombie virus in existence since 2010. You are a member of a small group of 225 individuals, 150 men and 75 women now residing in the deserted White House in Washington D.C. You are also the only member of the group with training in epidemiology. Twenty-five members of your group, all men, have developed a patch of blood red skin and are having difficulty seeing in the dark. People are frightened and there are many hypotheses: zombie molecules in the air, contaminated water, or the food that is being eaten. Rick Grimes, previously a police officer, said he believes the patchy skin is a result of minute zombie molecules floating in the air, while Daryl Dixon suspected it was from eating the fancy White House food, specifically the aged caviar that "ain't fit for no human." Reverend Gabriel Stokes whispered that he suspected it might be a sexually transmitted disease, and Carol Peletier, a savvy stay-at-home mom, said it could be prevented by eating her chocolate chip cookies as evidenced by the fact that Glen has been eating a chocolate chip cookie every day for the past three months and has avoided getting the skin patch. Coincidentally, Carol is offering to sell the recipe to her cookies for a bargain price of one extra food ration a day.

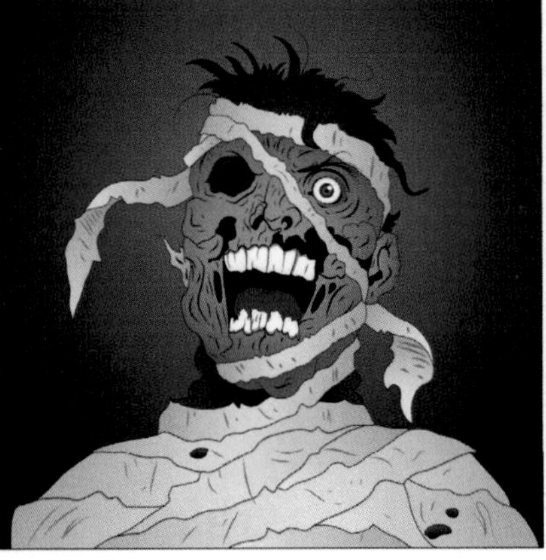

Given your training in epidemiology, everyone is looking to you to solve the mystery of this illness. One member of your group, Maggie Greene is a trained farm nurse, and another, Michonne, is a scientist who made quick use of the White House medical equipment, including a microscope. Michonne has set up in the laboratory facilities hidden in the underground White House bunker, while Maggie created a small clinic in the Oval Office. You remember Koch's postulates from an introductory epidemiology class and decide to use them to test the infectious disease hypothesis. You first instruct Maggie to draw blood from all of the individuals with the patch of blood red skin and provide them to Michonne to grow cultures. Michonne found that 75% of people with the patch have a positive culture for an unusual gram-positive bacteria, but she does not recognize the actual organism. Next you have Michonne inject bacteria from the culture into 15 male volunteers, 10 of whom then develop a patch of blood red skin in the first three days. Maggie then draws new blood samples from all the participants and Michonne set up new cultures, but none of these samples grew the gram-positive bacteria. Furthermore, five additional people who did not participate in the study developed symptoms.

Which of Koch's postulates are supported by your findings?

Table 1. Koch's postulates

Postulate	
1. The organism must be observed in every case of disease.	Y N
2. The organism must be isolated and grown in pure culture.	Y N
3. The pure culture must, when inoculated into a susceptible animal, reproduce the disease.	Y N
4. The organism, must be observed in, and recovered from, the experimental animal.	Y N

5. How convincing is the evidence that the disease is due to an infectious agent?

A. Totally convincing
B. Somewhat convincing
C. Not very convincing
D. Totally unconvincing

You then instruct Alice McDonald, a college student majoring in home economics to survey everyone in the group and identify their diet for the past two weeks. You provided her with an old epidemiology workbook with a list of guidelines for establishing causality, and tell her to include questions on the timing of events and the amount of caviar people ate.

Here is what she found:

Of the 40 men who are now identified with the mystery illness, 20 reported eating the aged caviar.
Of the 110 men without the illness, 50 also ate the aged caviar.
Alice calculated an odds ratio of 1.2 which was not statistically significant.
All of the 20 ill individuals who ate the aged caviar, stated they did so at least 2 days before they became ill.
There was no difference in the amount of caviar eaten by those with the rash and those without.

Since nearly everyone ate the caviar, it was not possible to see what would happen if people did not start eating the caviar. In addition, those who already ate the caviar stated it was absolutely addicting and they could not stop eating it.

Alice also reported that more people who ate the caviar complained of stomach upset and diarrhea than those who did not, and this occurred among both men and women.

Other items in the report. Alice stated she reviewed her nutritional textbooks and found an article describing a study in which mice developed a rash after being fed caviar. The researchers found that the rash increased the stomach acidity among the mice and they hypothesized that the caviar caused electrolyte imbalance that could lead to rash and respiratory distress. There were no studies listed describing other humans who had eaten the caviar but her resources were limited, due to the lack of individuals to manage the internet and online library databases from the flu pandemic.

Consider the causality guidelines described above and identify which of these guidelines support causality by eating the caviar. For each guideline, choose Yes (Y), No (N), or Not Tested (NT).

Table 2. Association of mystery illness with eating the caviar using the guidelines for causality.

6. Temporal relationship	Y N NT
7. Strength of the association	Y N NT
8. Dose-response	Y N NT
9. Replication of findings	Y N NT
10. Biologic plausibility	Y N NT
11. Consideration of alternative explanations	Y N NT
12. Cessation of exposure	Y N NT
13. Specificity of exposure	Y N NT
14. Consistency with other knowledge	Y N NT

15. How convincing is the evidence that the disease is due to eating caviar?

A. Strongly convincing
B. Somewhat convincing
C. Not very convincing
D. Totally unconvincing

Meanwhile, Fred Granite, an environmental engineer, stated that based on conversations he had with the people with the rash, a large number of them reported having drunk water from the cooler in the Oval Office. Furthermore some individuals reported there was a strange, somewhat sweet smell in water. Fred would like to explore this hypothesis. You develop a survey to identify the association of drinking water from the Oval Office with the mystery illness, and asked Fred to administer it. Here is what he found:

Of the men with the mystery illness, 25 reported drinking Oval Office water and 15 did not.
Of the men without the mystery illness, 61 reported drinking Oval Office water and 49 did not.
The odds ratio for this was 1.3, and it was not statistically significant.
No one who drank the Oval Office water reported any other illness, and all those ill reported drinking the water before becoming ill.
Consider the causality guidelines described above and identify which of these guidelines support causality by drinking the water from the Oval Office. For each guideline, choose Yes (Y), No (N), or Not Tested (NT).

Table 3. Association of mystery illness with drinking Oval Office water using the guidelines for causality

16. Temporal relationship	Y N NT
17. Strength of the association	Y N NT
18. Dose-response	Y N NT
19. Replication of findings	Y N NT
20. Biologic plausibility	Y N NT
21. Consideration of alternative explanations	Y N NT
22. Cessation of exposure	Y N NT
23. Specificity of exposure	Y N NT
24. Consistency with other knowledge	Y N NT

25. How convincing is the evidence that the disease is due to drinking water from the Oval Office cooler?

A. Strongly convincing
B. Somewhat convincing
C. Not very convincing
D. Totally unconvincing

People were getting even more concerned as the investigation lingered, although Carol was making a good profit with her cookie sales. You then remembered a lecture on the web of causation and examined your findings a bit further. You considered the possibility that this illness resulted from a number of factors. First you noted that only men acquired the illness so you suspected a genetic component. You then grouped men with the mystery illness by both exposures combined, eating caviar and drinking water from the Oval Office cooler", and this is what you found.

Of those with the mystery illness, 23 ate caviar and drank Oval Office water and 17 did not.
Of those without the mystery illness, 30 people ate caviar and drank Oval Office water while 80 did not.
The odds ratio was 3.6 and it was statistically significant.

26. Compile your results from the studies into the following table.

Table 4. Association of both risk factors with the mystery illness.

	Has mystery illness		Does not have mystery illness		Odds ratio
Exposure	Yes	No	Yes	No	
Ate caviar					
Drank water					
Ate caviar and drank water					

27. Using the concept of necessary and sufficient causes of disease described in number 5 previously, which of the four models best describes your results for eating caviar in Table 4?

A. Necessary and sufficient
B. Necessary but not sufficient
C. Sufficient but not necessary
D. Neither sufficient nor necessary

28. Using the concept of necessary and sufficient causes of disease described in number 5 previously, which of the four models best describes your results for drinking Oval Office water in Table 4?

A. Necessary and sufficient
B. Necessary but not sufficient
C. Sufficient but not necessary
D. Neither sufficient nor necessary

29. What public health warning will you give to the population?

A. Avoid eating the caviar
B. Avoid drinking water from the Oval Office
C. Avoid eating caviar and drinking the Oval Office water
D. Eat lots of Carol's cookies.

References

1. King LS. Dr. Koch's Postulates. J Hist Med. Autumn 1952:350-361.
2. Gordis L Epidemiology, 2nd Edition W.B. Saunders Company Philadelphia 2000.
   Disease

"Brain Boosting" using ADHD Drugs

Background

Ritalin, the trade name for methylphenidate, is a medication prescribed predominately for children with an abnormally high level of activity or with attention-deficit hyperactivity disorder (ADHD). Ritalin acts by stimulating the central nervous system, with effects similar to but less potent than amphetamines and more potent than caffeine. Ritalin has a notable calming effect on hyperactive children and a "focusing" effect on those with ADHD. When taken as prescribed, Ritalin is fairly safe, and funded research has shown that people with ADHD do not get addicted to Ritalin when used as prescribed. Adderall is also a psycho-stimulant medication indicated for use in ADHD. In the United States, both Ritalin and Adderall are Schedule II drugs under the Controlled Substance Act due to having significant abuse and addiction potential if not used correctly. Use of these drugs without a prescription is illegal and unauthorized use or sale of these drugs is a felony.

Controversy has arisen because people are taking Ritalin and Adderall as a "brain boost" to improve their ability to perform well on tests and other activities requiring intellectual concentration. In an editorial published in the journal, *Nature*, a number of scientists supported the idea that competent adults should be able to engage in what they refer to as "cognitive enhancement using drugs". The editorial goes on to call for further research into the risks and benefits of using drugs this way. There are other ways to improve concentration and intellectual ability without using drugs, such as getting enough sleep and learning new skills. Studies have found that between 5% and 15% of college students use brain-boosting drugs, mostly Ritalin or Adderall. These drugs can increase attention span, boost memory, and focus thinking; but they are not without risk. For example, people using these drugs without the supervision of a physician may have an increased risk of an overdose. Furthermore, there may be some risks if people use these drugs in conjunction with alcohol and other drugs. Approximately half of people who use these drugs as brain boosts report side effects; most commonly headaches, jitteriness, anxiety, and sleeplessness. A fair number of individuals discontinued using the drugs due to these side effects.

It is clear that increased research is necessary to identify the risks and effects of using ADHD drugs as a "brain boost" by individuals without ADHD or other illnesses treated by these medications. And since college students are a group who may use these drugs more frequently than others, as well are more likely to use alcohol and other drugs; research is needed to identify the risks in this population. And this is where epidemiology becomes important. In the following exercise, we will develop research studies to answer some of the important questions about the use of ADHD medications as brain boosters. The first exercise will demonstrate how a clinical trial would be conducted, and the second an observational study. Please note: these studies are hypothetical and the data provided are not real.

Sintix in College Students (SICS Study)

Dr. Raymond Romano is the Principle Investigator (person in charge of a study) of a randomized clinical trial to determine the effectiveness of Sintix, a new ADHD drug. He wants to determine if this drug improves examination grades among college students without ADHD. He is also going to evaluate the safety of this medication. Dr. Romano hired you to work as a research assistant on his study. Data from the study can be provided to the FDA to advise them on whether this medication actually does improve test scores among college students. Before you begin to work on this study, you should first read the following review of clinical trials.

Review of clinical trials:

A clinical trial is a research study in which two or more groups of individuals are randomly assigned to research groups, and followed over time to see if they develop an outcome. Randomized clinical trials generally include the following components:

1. Randomization: One group is given a treatment and the other is not. Generally people are randomly selected into a treatment and control group. This is called randomization or random assignment. The purpose for random selection is so that people are not chosen in a way that could bias a study. For example, a physician who is participating in a trial of a new treatment for asthma might believe the new drug is very good and only enroll his sickest patients in that part of the study. If this happened, even if the drug was effective it might look like it wasn't because the patients getting it were sicker to start with.
2. Placebo: Often (but not always) participants are either given the treatment or a placebo (which is a "fake pill or treatment"). Placebos are given because there is a very strong "placebo effect" in which if people believe they are getting a treatment they will feel better and often have measurable improvements in their health just due to their belief in the treatment. Sometimes placebos cannot be used because it is impractical or unethical.
3. Blinding: If a placebo is used, people are not informed of whether they are getting the real pill or the placebo. They are, however, told that they have a chance to get the real medication or a fake one.
4. Double blinding: In a clinical trial, the researchers administering the drug or evaluating outcomes also may not know which person is getting the treatment or the placebo. This is called double blinding. This is done so that a doctor or researcher won't record outcomes differently based on his/her knowledge of the drug. For example, if a patient, Phil Richards, is being treated by Dr. Mary Jones for depression and the doctor knows that he is getting the real drug, she may be more likely to think the patient is doing better than a patient who she knows is not getting the real drug. Double blinding is also used to prevent bias.

Hypothesis

The first step in creating a research study is to develop a hypothesis. The research hypothesis is very important as it helps you define exactly how you will conduct your study. Research proposals need to have clearly identified exposures and outcomes. Exposures are also called independent variables and outcomes are called dependent variables. Since you are performing a study to determine an effect, it is important that you can measure your exposures and outcomes. Dr. Romano asked you to assist by identifying the exposure and outcomes you would use, as well as look over some of the hypotheses he was considering and see which you felt would be the best.

1. Based on your reading in the background, which of the following would be your exposure?
   A. Better concentration
   B. Increased test scores
   C. Taking Sintix or a placebo
   D. Increased anxiety
   E. Increased nervousness
   F. Poor sleeping patterns

2. You also need to identify your outcome. In a clinical trial you can have more than one outcome. Which of the following would not be considered an outcome in the proposed study?
   A. Better concentration
   B. Increased test scores
   C. Taking Sintix or a placebo
   D. Increased anxiety
   E. Increased nervousness
   F. Poor sleeping patterns

3. There are several hypotheses to be tested in this study. Which of the following do you think is a well written research hypothesis for the proposed study?
   A. Sintix will be effective and safe for college students.
   B. Sintix can be used by college students in a safe manner.
   C. College students who use Sintix will have 50% higher scores on a standardized test than college students not receiving Sintix.
   D. College students receiving Sintix will have a harder time coping than college students not receiving Sintix.

Ethical issues

Since this is a clinical trial in which Dr. Romano will be testing a drug on college students, he is very concerned about potential ethical issues. He is in the process of developing a written informed consent form for the students to sign. The consent form needs to be written in an easy to read format at a reading level appropriate to the study subjects. Consent forms fall under the ethical principle of "respect for autonomy"; in which research subjects will understand the benefits and risks of participating in a research project, because they will be fully informed. Therefore, researchers have a responsibility to provide participants any important information about medications they may receive. Dr. Romano has asked you to review some of the potential things he should tell the participants about side effects of the medication. He made a list of potential side effects and other important information that he should tell the students, and he would like you to review it and let him know which items he should include. Select Y for those warnings you would include and N for those you would not include. You need to know the side effects of Sintix so look them up in order to complete this assignment. (Hint: You can use patient information inserts for Adderall as Sintix has the same properties (available online) and see Table 1 in the attached appendix.)

Table 1. List of potential items to be considered for inclusion in the informed consent.

Item	Include
4. Sintix may cause excessive anxiety.	Y N
5. Sintix may cause excessive sleepiness.	Y N
6. Students will be given either Sintix or a placebo and will not know which one they get.	Y N
7. Sintix impairs the ability of the users to engage in potentially hazardous activities; such as operating machinery or vehicles. They may not be able to drive when using it.	Y N
8. Sintix may be associated with birth defects if taken by pregnant women.	Y N
9. Sintix may cause increased blood pressure.	Y N
10. Sintix may cause a decrease in concentration.	Y N
11. Sintix may cause weight gain.	Y N

Selecting study subjects

The next thing Dr. Romano has to do is identify subjects for inclusion. By law, study participants need to be at least 18 to give informed consent or their parents would have to sign a consent form. Also, one cannot select subjects by gender, racial and ethnic group, or age unless there is a good reason. Dr. Romano made a list of potential criteria and he would like your assistance in determining if each one should be an inclusion (they are invited into the study) or an exclusion (they are not allowed into the study) criteria. You need to consider the ethical principle of justice, the safety of study subjects, and practical issues.

Please indicate which participants should be invited to participate in the research study. Circle "I" for included or "E" for excluded. This list will identify the potential subjects for the study. Following the identification of study subjects, you will randomize them to be in the treatment or placebo group.

Table 2. List of potential inclusion and exclusion criteria for the research study.

Criteria	
12. Students diagnosed with ADHD and currently taking Sintix	I E
13. Students taking at least 12 credits in the Fall semester	I E
14. Both male and female students	I E
15. Sexually active female students not using any contraception	I E
16. Any student aged 16-25	I E
17. Students with a history of drug abuse	I E
18. Pregnant students	I E
19. Only students at least age 18	I E
20. Spanish speaking students	I E
21. Students who live away from campus and only take online courses	I E
22. Students already using ADHD drugs to improve test scores	I E

The next step is to identify the actual study procedures that Dr. Romano would use in his study. He needs to be able to enroll students, obtain informed consent, collect baseline information and be sure the inclusion and exclusion criteria are met, and develop data collection procedures. Again, he made a list of study activities and asked you to review them and determine which would be the best order to conduct the study activities. Place a number from 1 (first activity) to 8 (last activity) in the column on the right side of Table 3.

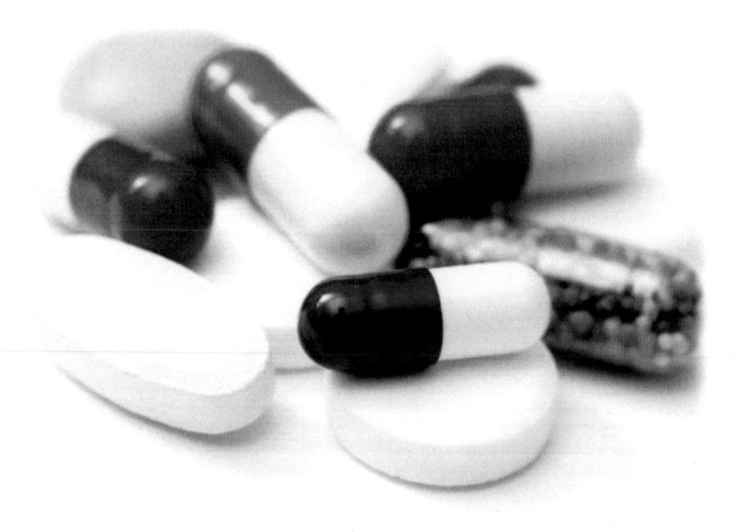

Table 3. Study activities

Activity	
23. Obtain written consent from all participants.	
24. Give participating students a test to determine their scores on a standard test of mathematical formulas.	
25. Inform students of which medication group they are in.	
26. Provide students with a study guide on mathematical formulas.	
27. Give a screening survey to identify inclusion and exclusion criteria	
28. Provide students with an envelope containing pills and instructions on the time they should take them.	
29. Survey the students to identify if they thought the medication they took had any side effects as well as whether they thought if it improved their test taking ability.	
30. Randomly assign students to the two study groups: Sintix and placebo	

31. In this study, neither the students nor the study coordinator knew which medication they were assigned. What kind of blinding was used in this study?
    A. No blinding
    B. Single Blinding
    C. Double Blinding
    D. Triple Blinding

The final list of students that Dr. Romano chose for his study were students enrolled in four randomly selected general education requirement classes who met the identified inclusion and exclusion criteria. You presented a description of the study to students at the beginning of each class session, telling them about the study, and informing them that students would be randomly assigned to one of two groups, those given Sintix and those given a placebo. Students would not be informed which group they would be assigned to, and each student who chose to participate would be given $50 to reimburse them for the time they provided. Each student was given a flyer about the study and told you would be available after class so they could set up an appointment with you to have a baseline screening performed and obtain informed consent. You knew that you needed at least 360 students in order to have enough power to identify a statistically significant difference in test scores between the two groups. You were able to enroll a total of 420 students; which would allow you to have enough in case any students dropped out of the study.

One way of placing people into study groups is to use a random numbers calculator. First, you would place the Sintix into an envelope labeled 1 and the placebo into an envelope labeled 2, or alternately you can use different colored envelopes. (In reality, you would probably use different numbers than 1 or

2 as people might guess that 1 would a treatment and 2 a placebo.) See http://stattrek.com/Tables/Random.aspx. Using this calculator, you indicate that you need 420 random numbers from a minimum of 1 to a maximum of 2. You will also need to enter true to allow for duplicate entries. If 1 = Sintix and 2 = Placebo, then as subjects are enrolled into the study, you read down the columns and determine which subject gets which treatment. In this example, the first person enrolled would get the placebo and the second person would get the Sintix. You would also need to create a database so that you would know which person in the study got which treatment. This database would be kept in a computer that was in a locked room with limited access. You would also print out a copy of the database in the event of computer failure as it would be very difficult if you needed to explain to Dr. Romano at the end of the study that you did not know who got which medication.

Table 4. 420 Random Numbers*

2 2 2 2 1 2 1 1 1 2 2 2 2 1 1 1 1 2 2 1 2 2 1 1 2 2 2 1 2 1 1 2 2 2 2 2 1 2 2 2 1 1 1 2 1 1 1 1 2 2 1 1 1 2 2 1 1 2
1 1 2 1 2 1 1 1 1 2 2 2 2 2 2 1 1 1 1 2 2 2 1 1 1 1 1 1 1 1 2 1 2 2 1 1 2 1 2 1 2 2 1 2 1 1 1 1 2 1 1 2 1 2 1 1 1 1
2 1 1 2 2 1 1 2 2 2 1 1 1 1 1 2 1 1 2 1 1 1 1 1 1 2 1 2 2 1 1 2 1 1 1 2 1 1 1 1 2 2 2 1 2 1 2 1 2 1 1 1 1 1 2 2 2 2
2 1 1 1 2 2 2 1 1 2 2 2 2 1 1 1 2 2 2 2 2 1 1 1 2 1 1 1 2 2 1 2 1 1 2 2 1 2 2 1 2 2 2 2 1 1 2 1 1 2 1 2 1 1 2 2 2 1
2 2 1 2 1 1 1 1 1 2 1 1 1 1 1 2 1 2 1 1 1 1 2 2 2 2 2 2 1 2 1 1 1 2 2 2 2 1 1 1 1 2 2 1 2 2 1 1 2 2 2 1 2 1 1 2 2 2
2 2 1 2 2 2 1 1 1 2 1 1 1 1 2 2 1 1 1 2 2 1 1 2 1 1 2 1 2 1 1 1 1 2 2 2 2 2 2 1 1 1 1 2 2 2 1 1 1 1 1 1 1 1 2 1 2 2
1 1 2 1 2 1 2 2 1 2 1 1 1 1 2 1 1 2 1 2 1 1 1 1 2 1 1 2 2 1 1 2 2 2 1 1 1 1 1 2 1 1 2 1 1 1 1 1 1 2 1 2 2 1 1 2 1 1
1 2 1 1 1 1 2 2 2 1 2 1 2 1

* This table of 420 random numbers was produced according to the following specifications: Numbers were randomly selected from within the range of 1 to 2. Duplicate numbers were allowed.

32. Using the random numbers generated in Table 4, how many of the first 10 subjects received the Sintix and how many received the placebo? Read across the Table to answer this question.

A. 2 Sintix and 8 Placebo
B. 5 Sintix and 5 PlaceboC.
C. 6 Placebo and 4 Sintix
D. 6 Sintex and 4 Placebo

33. What is the main reason study participants or individuals are randomly assigned to the study groups?
A. To create two identical groups
B. To keep the study subjects from knowing which group they are assigned to
C. To decrease the bias in how study subjects were assigned to groups
D. To prevent researchers from knowing which group study subjects were assigned to

At the end of a research study, the Principle Investigator will need to analyze the data and present the results of the study. This is done in a manner similar to the M&M assignment you completed earlier. Refer to that assignment if you are uncertain how to answer the next three questions.

34. In this **hypothetical** study Dr. Romano reported the following result, "This study found that taking Sintix resulted in higher scores on the math test, with an average score of 76 among students taking Sintix as compared to 68 among students not taking Sintix, with a p-value of 0.13." What should he conclude?

    A. Students on Sintix had significantly higher study scores.
    B. Students on Sintix had significantly lower test scores.
    C. There was no significant difference in test scores between those on Sintix and not on Sintix.

35. Dr. Romano also reported this second result, "This study found that college students taking Sintix were 2 times more likely to report difficulty sleeping than students not taking Sintix, with a p-value of 0.02." What should he conclude?

    A. Students on Sintix had significantly greater difficulty sleeping.
    B. Students not on Sintix had significantly greater difficulty sleeping.
    C. There was no significant difference in difficulty sleeping between those on Sintix and not on Sintix.

36. Based on the **hypothetical** results of this study, what would one conclude from the study?

    A. Sintix should be considered for use as a brain booster among college students; as it is useful in improving study scores with no significant side effects.
    B. Sintix might be considered for use as a brain booster among college students; as it is useful in improving study scores but has significant side effects.
    C. Sintix should not be considered for use as a brain booster among college students; as it is not useful in improving study scores and has significant side effects.
    D. Sintix should not be considered for use as a brain booster among college students; as it is not useful in improving study scores although it has no significant side effects.

The frequency of ADHD use as a Brain Boost among college students

Research studies can be useful in determining how frequently college students use ADHD drugs as a brain boost and the factors associated with its use. An epidemiologist, Dr. Emma Pillsbury hypothesized that students who live in dorms are more likely to use ADHD drugs as a brain boost than students who commute. She believes it is because students in dorms will be more exposed to other students using these medications. She also believes the use of these drugs will increase over time throughout the students' college careers. She is also concerned about the safety of students using these drugs in conjunction with alcohol and other drugs. Based on previous research she knows that about 10% of college students use ADHD drugs. Dr. Pillsbury develops a research study to study these issues. Based on the excellent job you did working for Dr. Romano, Dr. Pillsbury hired you as a research assistant on her project.

Dr. Pillsbury plans to enroll 2,000 newly admitted freshmen students living in dorms and 2,000 newly admitted freshmen students living off campus, and then interview them when they are first enrolled and twice every year until they graduate or drop out of school. She will conduct interviews in September and May.

37. What type of study is this?
    A. Case-control
    B. Cohort
    C. Cross-sectional
    D. Ecologic
    E. Clinical trial

38. What is the exposure in this study?
    A. Being a freshman
    B. Living in a dorm or off-campus
    C. Using ADHD drugs
    D. Drinking alcohol

39. What is the outcome?
    A. Being a freshman
    B. Living in a dorm or off-campus
    C. Using ADHD drugs
    D. Drinking alcohol

40. Dr. Pillsbury decided to only include students who were at least 18 years of age as they were able to consent for themselves. She is not sure what to do about students who already used ADHD drugs. What would you advise her?
    A. Include them because you want the students to be representative of all students, and she could create a bias by omitting them.
    B. Exclude them because they already have the outcome and in cohort studies, you need to start with people who are free of the outcome.

In this **hypothetical** study, Dr. Pillsbury found the following results:

1. 13% of college students used ADHD drugs, with 15% in their freshman and sophomore years and dropping to 10% in their junior year and 6% in their senior year.
2. Students who took ADHD drugs had a higher GPA than those who did not use these drugs, 3.15 vs. 2.95 respectively.
3. Students who used ADHD drugs were more likely to abuse alcohol than students who did not use these drugs, 18% vs. 12% respectively.
4. Students who used ADHD as freshmen were more likely to drop out than students who did not use these drugs, 10% vs. 7% respectively.

Comparison between study designs

There are advantages to each study design. In the first design, the randomized clinical trial, the researchers has more control over the study subjects and what they will do, while in the cohort study (I just gave you the answer to question 37, by the way.), the researcher observes what students are doing.

41. Both studies found that students who took ADHD drugs had higher test scores. Can Dr. Pillsbury report that taking ADHD results in a higher GPA as Dr. Romano was able to state for the math score in the clinical trial?
    A. Yes, because the GPA was higher among students who used ADHD drugs.
    B. No, because even though their GPA was higher, the students who used ADHD drugs may have been more motivated to study and care more about their GPA than those who did not take ADHD drugs.
    C. No, because even though their GPA was higher, the students who used ADHD drugs might have already been higher achieving students when they enrolled in the university.
    D. Both B and C

42. Can Dr. Romano state that his study showed that 50% of college students used Sintix since that is the number of students who used the drug in his study?
    A. Yes because 50% of the subjects in his study used Sintix.
    B. No, because he is only measuring the number of students who he gave the drug to and not the number who used the drug on their own.

Thus, you can see that different study designs answer different questions. When a researcher wants to study the effects of a medication, a randomized clinical trial is more effective, but if a researcher wants to understand how many students use a medication and the types of behavior associated with that behavior, an observational study, like a cohort study would be more effective. Again, I want to remind you that these data were used as an example and no studies as described in these exercises were actually conducted using Sintix as a brain booster. However, these types of studies would be important to do so we can understand how students use ADHD drugs and the potential benefits and risks of doing so.

Sources
http://www.drugfree.org/Portal/Drug_guide/Ritalin
http://en.wikipedia.org/wiki/Ritalin
http://en.wikipedia.org/wiki/Adderall
http://www.webmd.com/brain/news/20081211/brain-boosting-faq-what-you-must-knowhttp://stattrek.com/Tables/Random.aspx

Table 5. Adverse Events Reported by 5% or More of Adults Receiving Sintix with Higher Incidence Than on Placebo in a 500 Patient Clinical Trial Study*

Body System	Preferred Term	Sintix® (n=191)	Placebo (n=64)
General	Asthenia	6%	5%
	Headache	26%	13%
Digestive System	Loss of Appetite	33%	3%
	Diarrhea	6%	0%
	Dry Mouth	35%	5%
	Nausea	8%	3%
Nervous System	Agitation	8%	5%
	Anxiety	8%	5%
	Dizziness	7%	0%
	Insomnia	27%	13%
Cardiovascular System	Tachycardia	6%	3%
Metabolic/Nutritional	Weight Loss	11%	0%
Urogenital System	Urinary Tract Infection	5%	0%

Note: The following events did not meet the criterion for inclusion in Table 3 but were reported by 2% to 4% of adult patients receiving Sintix with a higher incidence than patients receiving placebo in this study: infection, photosensitivity reaction, constipation, tooth disorder, emotional lability, libido decreased, somnolence, speech disorder, palpitation, twitching, dyspnea, sweating, dysmenorrhea, and impotence.

*Included doses up to 60 mg.

Module 7 – Health Policy and Communication

Much of health policy is dependent on effective communication. In order to convince the public why they should follow and accept the policy, public health advocates must communicate with the public. This assignment will focus on a hot-button issue in public health today: vaccination of children. You are the public health consultant for a county school district and have been tasked with informing the parents of the incoming students about the vaccination policy. Write a communication (memo) informing parents of vaccination policy and procedures. Be sure to cover

- who (must be vaccinated prior to entering school)
- what (proof must be provided)
- where (they can obtain services)
- why (convince your audience – what will happen if they do not vaccinate?)

Remember, be creative in your policy! This does not need to be representative of the current policy, but what you feel the policy should or could be. The websites listed below are a good source of information to familiarize yourself with vaccination policy information.

- http://www.cdc.gov/vaccines/imz-managers/laws/state-reqs.html
- http://www.theatlantic.com/education/archive/2015/02/schools-may-solve-the-anti-vaccine-parenting-deadlock/385208/
- http://vaccines.procon.org/
- http://www.ncbi.nlm.nih.gov/pmc/articles/PMC2553651/

Before you begin, please review the information on memo writing so that you can write an effective memo. There is also an excellent source of information on memo writing at this website: http://owl.english.purdue.edu/owl/resource/590/1/

Memo Writing

There are three reasons for writing a memo: the first is to bring someone's attention to a problem, the second is to solve a problem, and the third is to encourage action on the recipient's part. Memos are formal documents and need to be written in a professional manner. Writing a memo is different than writing a term paper, both in structure and in type of language used. Memos are read by busy individuals, and they do not want long background sections and flowery language. Memos are short and very specific. The purpose of a memo is often to persuade an individual to do something in a short, concise message.

The person writing a memo has a number of responsibilities as the recipient may act based upon the information provided. In many instances, memos are written to one's boss or superior. If a superior acts upon a memo sent that is based in error or does not consider the consequences of the action recommended, the memo author's career may be at risk and harm may be done to his or her constituents. In other instances, inner-office memos can be released to the public, and if not written effectively, can destroy the credibility of the office. Thus, the writer needs to provide a short and clear analysis of a situation that is both balanced and neutral, and the writer need to consider the advantages and disadvantages of any recommended action before making recommendations in writing.

Although memos are short, usually no more than 2-3 pages, they need to address the key issues, provide a framework for understanding these issues, and include a specific recommendation for taking action. The writer needs to consider the audience and not make recommendations beyond what that person can do. For example, it would be foolish to recommend that a parent change the district's vaccination policy if they do not like it as they would lack the power and ability to do so. However, it would be reasonable to provide information on how to lodge a complaint if the addressee disagreed with the suggested action. Thus, the memo writer needs to clearly consider the recipient of the memo, and the practicalities of any recommendations.

There are distinct parts to a memo. The first is the heading which identifies who wrote the memo, who it was sent to, the date, and the subject.

Heading

TO: Recipient's name and title
FROM: Your name and title
DATE: Current Date
SUBJECT: The purpose of the memo

Body of the Memo

This section provides a description of the purpose of the memo, and includes any important background information, as well as the specific recommended response. You should describe the background as succinctly as possible, but provide sufficient detail. The challenge is to provide only as much information as is necessary, but enough detail so the recipient can understand the issue and suggested action. You need to close the memo with a polite ending that clearly indicates what action you want the recipient to take.

Attachments

You can attach any documents needed to provide detailed information that the recipient may need. Be sure to identified any attachments in your memo under the closing

Attached: List the name of the attachment

Disease screening

1. Screening for disease is conducted so that a disease can be identified in the early phase and early treatment can be initiated to prevent subsequent mortality from the disease. Thinking back to the information on prevention, what type of prevention occurs when screening identifies a disease early in the process?
   A. Primary
   B. Secondary
   C. Tertiary

There are different screening techniques, one involves screening everyone for disease and the other involves screening high risk individuals for a disease. There are advantages and disadvantages of each approach which is affected by the sensitivity, specificity, positive and negative predictive value of tests. Let's review these concepts.

Sensitivity refers to the number of individuals with a disease who correctly test positive for the disease. Specificity refers to the number of people without a disease who correctly test negative for the disease. Positive predictive value refers to the number of individuals who test positive and do have the disease. Negative predictive value refers to the number of individuals who test negative and do not have the disease.

Look at the 2x2 table presented below and write the letter corresponding to the correct formula for each component of screening on the line to the right of the word.

	Disease		
Test Results	**Yes**	**No**	
Positive	a	b	a+b
Negative	c	d	c+d
	a+c	b+d	

A: a/(a+b)

C. a/(a+c)

B. d/(b+d)

D. d/(c+d)

2. Sensitivity ________
3. Specificity ________
4. Positive Predictive Value ________
5. Negative Predictive Value ________

Which letter in the table refers to the each of the following?

6. True Positives ________

7. False Positives ________

8. True Negatives ________

9. False Negatives ________

Sensitivity and specificity are not affected by the prevalence of the condition, as sensitivity only evaluates individuals with a disease while specificity only evaluates individuals without a disease. However, positive predictive value and negative predictive value are affected by the prevalence of a disease. The relationship between positive predictive value and prevalence, and how it can impact on screening decisions is illustrated in the exercise below.

Screening for Diabetes

You have just been hired to conduct a diabetes prevention program for a local community advocacy program, and you have heard there are some strong personalities among board members. A local laboratory has developed a quick screening test for diabetes and they have donated 20,000 test kits to your project. They hope that a positive reaction from the community to their test will help their image, and ultimately lead to sales of their screening test. According to Mark Philips, CEO of the company, the test has a sensitivity of 99% and a specificity of 95%. You have scheduled a meeting with your board to identify the community in which to conduct the screening tests. Prior to the meeting Mark raised two concerns; the first is that he would like you to use the test to find the highest possible percent of diabetic patients, and secondly he is concerned about the community perception of this test. He whispered that he was particularly worried about false positives as he knew people who tested positive would have to take time off from work for additional screening and he was worried about the negative publicity from that, especially in light of the fact that he would not be able to cover the costs of any additional screening (estimated at $50 per person).

You met with your board on Monday morning and it was apparent from the start that there were some strong personalities. Mary Lou Andres began by complaining that as a diabetes prevention program, they should not be serving donuts. Paula Marshall however, was happily claiming the chocolate frosted ones were the best in the county, while placing two of them at her spot on the table. You take a deep breath and suggest everyone gets down to business.

You are happy to inform the board of Mark Phillip's generous gift, and state that you want to maximize the effectiveness of the screening test. Mary Lou immediately demands that the organization conduct mass screenings of the entire community as she doesn't want any group to feel discriminated against. As she states, "It is the only fair way to conduct the screening". But Paula vehemently disagrees. She feels you should conduct selective screenings among individuals who are obese because they are at high risk for diabetes, and you would get "more bang for your buck". Mary Lou responded by telling Paula that she just doesn't know the community and they are headed for chaos if they follow her plan. Paula slams down her books and tells Mary Lou to get off her high horse. At this point you call for a coffee break and tell the board to have a few more donuts and then you will present some information that might help them decide the best course of action.

You pull out your calculator and munch on a maple walnut donut. You know from past experience that the prevalence of diabetes is 2% in the overall community and 20% in the obese group. You know if you complete the two following tables you can demonstrate how the prevalence rates of diabetes will affect your study results.

Table 1 shows what would happen if you follow Mary Lou's advice and conduct the study in the overall community. As we noted previously, the prevalence of diabetes is 2% in this community. Fill in the numbers of people with and without the disease based upon this prevalence. Then, using a sensitivity of 99% and specificity of 95%, fill in the 4 cells of the table identifying the number testing positive and negative for those with and without the disease. Identify the total numbers who test positive and negative and calculate the positive predictive value.
Hint: This table can be completed thinking logically and it is not difficult once you do it. Start by identifying how many have and do not have the disease. This is done by using the prevalence. Then when you have the bottom numbers figured out, use the sensitivity value to fill in for those who have the disease and the specificity value to fill in those who don't have the disease. You now have the 2 x 2 table completed so you can use the formula to calculate the positive predictive value.

Table 1. Calculations for diabetes prevalence of 2%

	Disease		Totals
Test	Yes	No	
Positive			
Negative			
			10,000

10. What is the positive predictive value for the low risk community?
    A. 29%
    B. 71%
    C. 95%
    D. 99%

11. In the community group what would be the total cost of the additional medical screening for each person who tests positive?
    A. $15,000
    B. $22,180
    C. $34,400
    D. $68,800

12. What would be the cost for each for each positive case identified? ______
    Hint: This is the total cost of additional screening/number of true positives

Table 2 information: prevalence of 20%, sensitivity of 99% and specificity of 95%. Calculate the positive predictive value.

Table 2. Calculations for diabetes prevalence of 20%

	Disease		Totals
Test	Yes	No	
Positive			
Negative			
			10,000

13. What is the positive predictive value for the obese individuals?
    A. 17%
    B. 83%
    C. 95% demonstrates what would happen if you conducted the testing among the obese community who would have a prevalence of diabetes of 20%, as suggested by Paula. Fill in all the parts of the table using the following
    D. 99%

14. In the community group what would be the total cost of the screening each person who test positive?
    A. $22,180
    B. $54,800
    C. $119,000
    D. $238,000

15. What would be the cost for each for each positive case identified? ___

16. Given Mark Phillips concerns, what screening strategy would you recommend?
    A. Targeted screening of obese individuals
    B. Mass screening of the entire community

Module 8 – Outbreak Investigation

One important public health service epidemiologists provide is assisting in an investigation of an outbreak. There are a variety of outbreaks epidemiologists may investigate including infectious disease, food poisoning, or environmentally-caused diseases.

Why should we investigate outbreaks?

There are many reasons why we must investigate outbreaks, rather than just quarantine entire cities like Europe did during the plague outbreak. While quarantining is certainly a viable option in extreme cases, there are often better and less intrusive methods to stop an epidemic. Some of the important reasons are:

- In order to control and stop the outbreak from spreading, one must identify the source
- It is important to investigate and identify strategies to prevent future outbreaks
- We may learn more about known diseases and their transmission
- It is also possible to identify and describe new diseases
- Part of an epidemiologist's job is to help protect the public and this includes stopping a panic (3)

Now that we know why we investigate outbreaks, how do we know when an investigation is warranted? There are many important factors to consider: Is there ongoing illness and exposure with new people continuing to get sick? How severe is the illness? How easy is it transmitted? How much is the public concerned? Sometimes a small outbreak will be investigated if only to calm the general public. While that may not seem to make sense, part of the primary goal of an outbreak investigation is to keep the public safe, both from real and imagined dangers. Once you have answered these questions, you will know whether an investigation (and the time and cost involved) is justified.

If an outbreak investigation is to be conducted, it is of vital importance to be systematic and meticulous in order for the investigation to be effective. Every person participating in the investigation should follow the same steps, ask the same questions, and gather comparable data. Coordination is key! It is also important to realize that an outbreak investigation is a moving, evolving system. Epidemiologists should stop often to assess what they have learned and adjust their investigation accordingly. An outbreak investigation is unique in that often where the investigation ends up is not where it started at all. It is imperative that epidemiologists and other health professionals generate hypotheses as new information is learned. Only in this way can cases be properly identified, leading to eventual control of the outbreak.

For fun ☺: Visit the website http://www.healthline.com/health/worst-disease-outbreaks-history#Overview1 to learn about the 10 worst outbreaks in U.S. history. Do you think these outbreaks could happen again in today's world?

What are the steps to use in investigating outbreaks? The CDC has identified the steps for investigation a (foodborne) outbreak, but these steps can easily be adjusted to investigate other types of outbreaks. (4)

1. Detecting a Possible Outbreak
    - Determine if there is an increased number of cases or clustering of cases
    - Review previous cases of the diseases or illness

2. Defining and Finding Cases
    - Clinically describe the cases as clearly and with as much detail as is known
    - When possible, utilize the "time, place, and person" case definition identifying the time of illness, the place illness began, and the person who became ill
    - Use this description to find similar cases, even those that may not have been initially identified as a case

3. Generating Hypotheses about Likely Sources
    - Using gathered information, describe the outbreak over time (can be hours, days, weeks, etc. depending on the type of outbreak)
    - Use epidemiological tools (i.e. epidemic curve, spot map, incidence mapping, case interviews) to try to determine commonalities among the cases

4. Testing the Hypotheses
    - Design and implement an epidemiological study (often case-control) to test your hypotheses
    - Analyze the data gathered during the study
    - Draw conclusions that can explain the facts gathered and observed

5. Finding the Point of Contamination
    - While not always possible (i.e. ebola), when possible locate the point of contamination
    - Test your initial identification of the point of contamination against the facts gathered during interviews and your epidemiological study

6. Controlling an Outbreak
    - Once the facts are gathered and confirmed action must take place
    - How the outbreak is controlled will be dictated by the type of contamination
        - For example: environmental – blocking access to an area, food – removing the food source, infectious disease – isolating the infected
    - Controlling an outbreak often requires significant coordination and cooperation between public and private industries, the government, and the general public
        - Effective communication often makes the difference between success or failure in controlling an outbreak

7. Deciding an Outbreak is Over
    - Implement a protocol to determine when an outbreak is over
    - Develop a contingency plan to ensure any future cases or outbreaks are identified and controlled quickly
    - Communicate your findings, new information learned, and status of the outbreak

TO DO:

1. Read the case report on the 2015 salmonella outbreak http://www.cdc.gov/salmonella/live-poultry-07-15/

2. Read the article on effective questioning of potential outbreak cases (http://nciph.sph.unc.edu/focus/vol2/issue1/2-1HypInterviews_issue.pdf) and use the information to complete the questionnaire assignment explained below

Case Interviews: A potluck gone wrong

"When an outbreak occurs, one of the first tasks is to interview case-patients and health care providers. These interviews give initial clues to possible sources of exposure. In addition, the information from these initial interviews can be used to develop another, more detailed questionnaire for testing the hypothesis generated through initial interviews." (1)

Every year the residents of Temple Terrace, Florida look forward the town barbeque and fireworks on the 4th of July at the golf course. It is a chance for neighbors get together in a way they don't usually have time to. Children and pets are seen running freely and everyone enjoys the delicious food. There are usually 5-10 barbeques fired up at the same time, along with endless tables of salads, vegetable dips, side dishes, and desserts. It is a great time of community togetherness that everyone enjoys.

On the morning of July 5th, PTA President Nancy Snow woke up planning to go to the annual debriefing meeting of the town festivities, but instead found herself feeling very ill. She had a fever, vomiting, and general malaise. She called her friend Susan Thomas who she was supposed to ride to the meeting with to cancel. To her surprise, Susan told her that she had been up most of the night with her 10-year old son who was vomiting also. Nancy wondered if it was coincidence or "something going around." Nancy then sent out a group text to the planning committee to let them know that neither she nor Susan would be able to attend due to illness. As the replies began coming through, it became obvious that something was wrong. Six of the ten members of the committee were either sick or someone in their family was suffering from the same symptoms as Nancy and Susan's son. Fearing an outbreak of some kind, Nancy had her husband call the local public health office to inform them.

You are a newly hired epidemiologist in the Temple Terrace Public Health Office. Your boss tasks with you with the job of interviewing the potential cases and reporting back to him the information. Given the little direction he provides, you start to panic, but then you remember the information you learned in your Intro to Epidemiology course and get to work writing an effective questionnaire.

Assignment: Write a 20 question interview form that you would use to interview potential cases in Temple Terrace. Remember to include questions that gather information about

- **Basic demographic information**
- **Clinical details about their symptoms**
- **Information about their activities at the barbeque**
- **Details on food consumption**

Sources

(1)http://nciph.sph.unc.edu/focus/vol2/issue1/2-1HypInterviews_issue.pdf
http://sphweb.bumc.bu.edu/otlt/MPH-Modules/EP/EP713_DescriptiveEpi/EP713_DescriptiveEpi3.html
http://cphp.sph.unc.edu/focus/vol2/issue1/2-1HypInterviews_issue.pdf
(3)http://epi.publichealth.nc.gov/cd/lhds/manuals/cd/training/Module_1_1.6_ppt_OutbreakInvestigation.pdf
(4)http://www.cdc.gov/foodsafety/outbreaks/investigating-outbreaks/investigations/
https://wiki.ecdc.europa.eu/fem/w/wiki/671

Complete the puzzle.

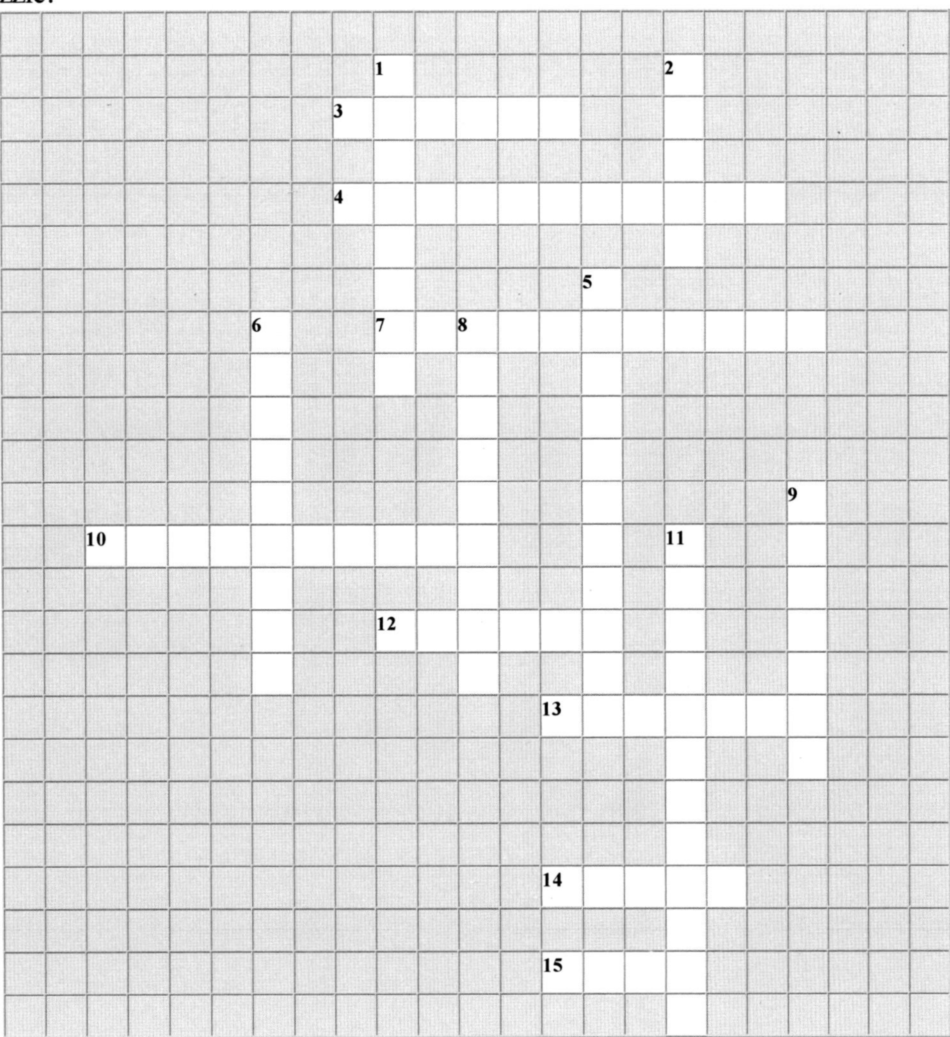

Down

1. Infections transmitted from vertebrate animals to humans.
2. C. botulinum can produce this type of substance.
5. A common source epidemic where all the cases were infected at the same time.
6. The first case of Hanta that comes to investigators attention in an epidemiologic investigation is an example of being a(an) ______ .
8. The strength of the pathogen to cause clinical symptoms.
9. Infectious individual who has the disease, but has no symptoms.
11. Protection against an epidemic that can be achieved when a certain level of people in a population get vaccinated for a disease.

Across

3. Doorknobs are an example of this type of vehicle.
4. Ability of an agent to get inside the body and replicate itself
7. Part of the epidemiologic triangle that refers to an area outside of the body where the disease, such as a bacterium may live.
10. (ill/(ill+well)) x 100 in a time period
12. A mosquito or any other insect that can transmit a pathogen.
13. Non-moving objects like food and water that by having contact with it can transmit a pathogen.
14. Part of the epidemiologic triangle that refers to a biologic, chemical, or physical hazard essential to causing a disease.
15. Part of the epidemiologic triangle that refers to a person who has the disease.

WORKBOOK ANSWERS

Module 1: Crossword Puzzle

Answer Key

1 P
2 J A
3 E X P O S U R 4 E 5 H 6 P O P U L A 7 T I O N
H P I E D
N I 8 P R I M A R Y R E
G D P 9 O U T C O M E
R 10 S E C O N D A R Y I I
A M C A C
U I 11 R I S K F A C T O R 12 M
N C A Y O
T T 13 A R
14 D E T E R M I N A N T B
S A I
15 M O R T A L I T Y D
Y I
T T
16 E P I D E M I O L O G Y
T C
17 J O H N S N O W
I
18 D E S C R I P T I V E
S

Module 2: Answers to Ethics questions

#1. Respect for autonomy. The researcher is not giving individuals the opportunity to decide if they wish to be in a study. One could also argue non-maleficence as the release of personal information could be harmful to the individual.

#2. Respect for autonomy. The rewards of participating in this study are so great that the prisoners may feel coerced to participate even if they do not really want to. This is why we are very careful in conducting research on prisoners who have less freedom to refuse.

#3. Justice. The drug is very expensive and it is quite likely that the population it is used on would not have the resources to ultimately receive the medication if it proved successful. One could also consider respect for autonomy as the participants have so few other resources that they will engage in this study as their only hope to receive medication.

#4. Beneficence. One always worries when conducting a randomized trial if one is doing good for the participants. In this case since we really do not know if the supplement is more effective than the existing treatment, it is ethical to conduct a trial. One does have to take into account special protections when conducting a study among children.

#5. Justice. Clearly only individuals with health insurance can benefit from this trial, as those without cannot. It is often true that individuals without health insurance or significant financial resources have limited ability to take part in these more expensive clinical trials.

Module 3: M&M Assignment

Example of the format of the answers. Your actual answers will differ.

A. First make a bar graph and paste it in here.

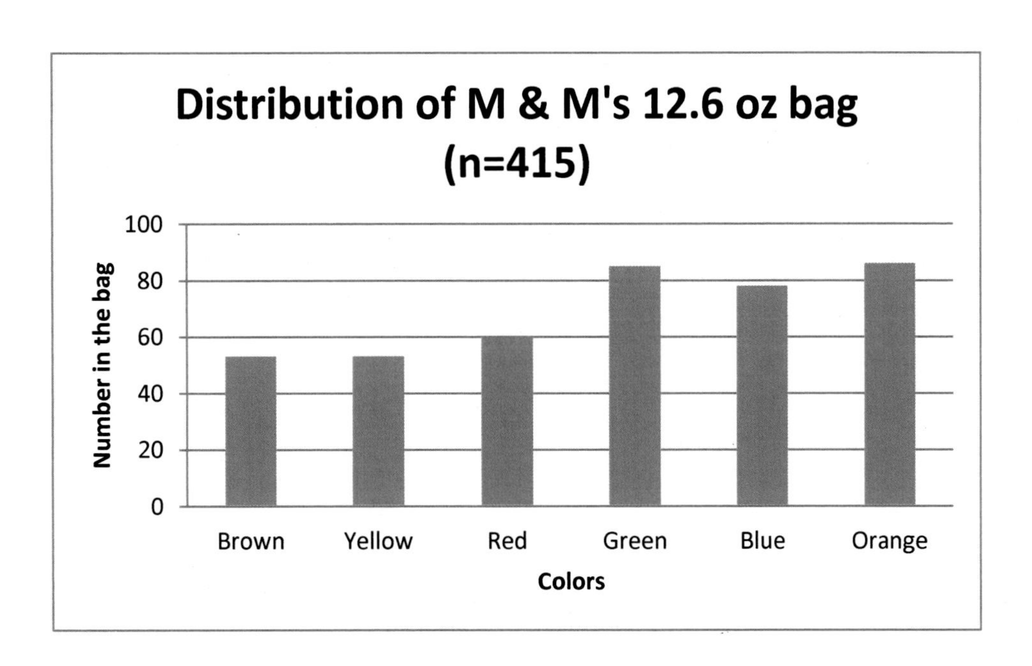

B. Now make a pie chart and paste it in here. Again be sure that you label the sections and describe what is in the figure.

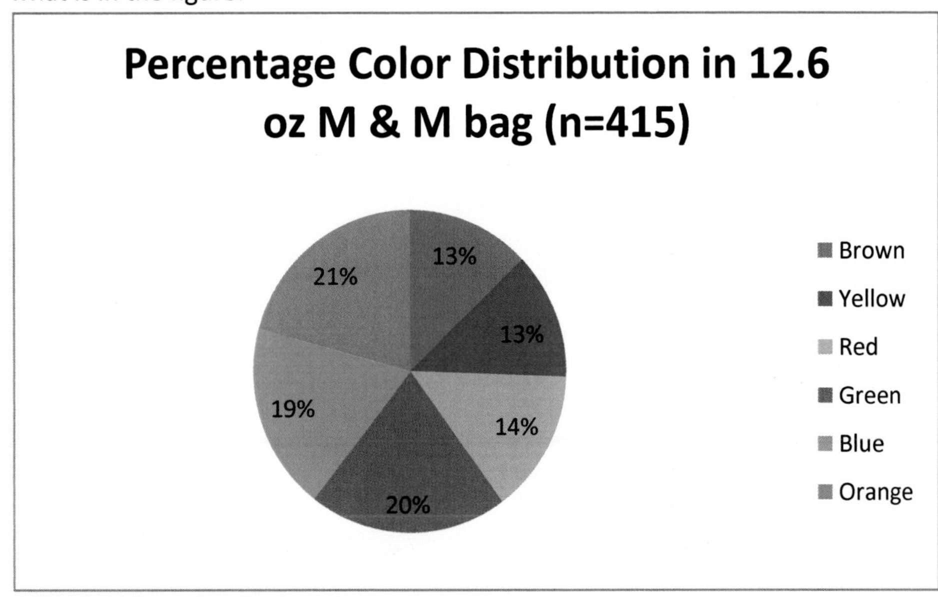

C. Write down the ratio of blue to red M&Ms.

Your answer should look this: 53/60

D. Write the proportion of each color using two decimal points.

Brown .13

Yellow .13

(Continue on for each color)

E. Show the percentage of yellow M&Ms. Be sure to include the % sign.

13%

F. Calculate the rate of brown M&Ms that you obtained at the start of this exercise. Include the time that you identified these M&Ms, e.g., on March 1st at 5pm.

12.8 per 100 M&Ms on November 1, 2009.

G. Now calculate the rate of the brown M&Ms that remain at this point.

14.8 per 100 M&Ms on November 1, 2009.

Statistical Chi Square Analysis

H. Write a null hypothesis below about what you expect to find.

Ho: There is no difference in the percentage of M&Ms in my bag as compared with the percentage presented by the MARS company.

The rest of the assignment is to be completed in Canvas.

Module 4: Rates

MATCHING

Rates

1. Crude death rates ___e___
2. Case fatality rate (%) ___i___
3. Cause specific rate ___f___
4. Age-specific rate ___j___
5. Sex-specific rate ___h___
6. Maternal mortality rate ___d___
7. Fatal death rate ___c___
8. Late fetal death rate ___k___
9. Crude birth rate ___b___
10. General fertility rate ___g___
11. Perinatal mortality rate ___a___

Formulas

(a) $\dfrac{\text{Number of late fetal deaths after 28 weeks Gestation + infant deaths within 7 days of birth}}{\text{Number of live births + number of late fetal deaths}} \text{ X 1,000}$

(b) $\dfrac{\text{Number of live births within a given period}}{\text{Population size at midpoint during that period}} \text{ X 1,000}$

(c) $\dfrac{\text{Number of fetal deaths after 20 weeks gestation}}{\text{Number of live births + number of fetal deaths after 20 weeks or more gestation}} \text{ X 1,000}$

(d) $\dfrac{\text{Number of deaths from childbirth}}{\text{Number of live births}} \text{ X 100,000}$ live births (during a year)

(e) $\dfrac{\text{Number of deaths in a given year}}{\text{Reference population (during midpoint of the year)}} \text{ X 100,000}$

(f) $\dfrac{\text{Mortality (or frequency of a given disease)}}{\text{Population size at midpoint of time period}} \text{ X 100,000}$

(g) $\dfrac{\text{Number of live births within a year}}{\text{Number of women aged 15-44 years during the midpoint of that year}} \text{ X 1,000 women}$

(h) $\dfrac{\text{Number of deaths among men}}{\text{Number of men (during time period)}} \text{ X 100,000}$

(i) $\dfrac{\text{Number of deaths due to disease “X”}}{\text{Number of cases of disease “X”}} \text{ X 100}$ during a time period

(j) $\dfrac{\text{Number of deaths among individuals aged 15-35 years}}{\text{Number of persons aged 15-35 years (during time period)}} \text{ X 100,000}$

(k) $\dfrac{\text{Number of fetal deaths after 28 weeks gestation}}{\text{Number of live births + number of fetal deaths after 28 weeks or more gestation}} \text{ X1,000}$

U.S. STANDARD CERTIFICATE OF DEATH

LOCAL FILE NO. STATE FILE NO.

NAME OF DECEDENT ____ For use by physician or institution

To Be Completed/ Verified By: FUNERAL DIRECTOR:

1. DECEDENT'S LEGAL NAME (Include AKA's if any) (First, Middle, Last): Joffrey Baratheon
2. SEX: Male
3. SOCIAL SECURITY NUMBER: unknown

4a. AGE-Last Birthday (Years): 19
4b. UNDER 1 YEAR: Months / Days
4c. UNDER 1 DAY: Hours / Minutes
5. DATE OF BIRTH (Mo/Day/Yr): 2/16/1448
6. BIRTHPLACE (City and State or Foreign Country): King's Landing

7a. RESIDENCE-STATE
7b. COUNTY: Seven Kingdoms
7c. CITY OR TOWN: King's landing
7d. STREET AND NUMBER
7e. APT. NO.
7f. ZIP CODE
7g. INSIDE CITY LIMITS? ☐ Yes ☒ No

8. EVER IN US ARMED FORCES? ☒ Yes ☐ No
9. MARITAL STATUS AT TIME OF DEATH: ☒ Married ☐ Married, but separated ☐ Widowed ☐ Divorced ☐ Never Married ☐ Unknown
10. SURVIVING SPOUSE'S NAME (If wife, give name prior to first marriage): Margaery Tyrell Baratheon

11. FATHER'S NAME (First, Middle, Last): Robert Baratheon
12. MOTHER'S NAME PRIOR TO FIRST MARRIAGE (First, Middle, Last): Cersei Lannister

13a. INFORMANT'S NAME: Mrs. Baratheon
13b. RELATIONSHIP TO DECEDENT: wife
13c. MAILING ADDRESS (Street and Number, City, State, Zip Code): King's Landing

14. PLACE OF DEATH (Check only one: see instructions)
IF DEATH OCCURRED IN A HOSPITAL: ☐ Inpatient ☐ Emergency Room/Outpatient ☐ Dead on Arrival
IF DEATH OCCURRED SOMEWHERE OTHER THAN A HOSPITAL: ☐ Hospice facility ☐ Nursing home/Long term care facility ☐ Decedent's home ☒ Other (Specify): Restaurant

15. FACILITY NAME (If not institution, give street & number): Red Keep
16. CITY OR TOWN, STATE, AND ZIP CODE: King's Landing
17. COUNTY OF DEATH: Seven Kingdoms

18. METHOD OF DISPOSITION: ☐ Burial ☐ Cremation ☐ Donation ☒ Entombment ☐ Removal from State ☐ Other (Specify):
19. PLACE OF DISPOSITION (Name of cemetery, crematory, other place): Lannister Family Tombs

20. LOCATION-CITY, TOWN, AND STATE: Dubrovnik
21. NAME AND COMPLETE ADDRESS OF FUNERAL FACILITY: Littlefinger Funeral Home, Dubrovnik

22. SIGNATURE OF FUNERAL SERVICE LICENSEE OR OTHER AGENT: George R. Martin
23. LICENSE NUMBER (Of Licensee): 001

ITEMS 24-28 MUST BE COMPLETED BY PERSON WHO PRONOUNCES OR CERTIFIES DEATH
24. DATE PRONOUNCED DEAD (Mo/Day/Yr): 4/13/1467
25. TIME PRONOUNCED DEAD: Unknown

26. SIGNATURE OF PERSON PRONOUNCING DEATH (Only when applicable): Grand Maester Pyceelle
27. LICENSE NUMBER: 007
28. DATE SIGNED (Mo/Day/Yr): 4/13/1467

29. ACTUAL OR PRESUMED DATE OF DEATH (Mo/Day/Yr) (Spell Month): 04/13/1467 April
30. ACTUAL OR PRESUMED TIME OF DEATH: unknown
31. WAS MEDICAL EXAMINER OR CORONER CONTACTED? ☐ Yes ☒ No

CAUSE OF DEATH (See instructions and examples)

32. **PART I.** Enter the chain of events--diseases, injuries, or complications--that directly caused the death. DO NOT enter terminal events such as cardiac arrest, respiratory arrest, or ventricular fibrillation without showing the etiology. DO NOT ABBREVIATE. Enter only one cause on a line. Add additional lines if necessary.

Approximate interval: Onset to death

IMMEDIATE CAUSE (Final disease or condition ---------> resulting in death)
a. Acute cyanide poisoning
Due to (or as a consequence of):

Sequentially list conditions, if any, leading to the cause listed on line a. Enter the **UNDERLYING CAUSE** (disease or injury that initiated the events resulting in death) **LAST**
b. Poisoned wine at a wedding
Due to (or as a consequence of):
c. ____
Due to (or as a consequence of):
d. ____

PART II. Enter other significant conditions contributing to death but not resulting in the underlying cause given in PART I: Evil lifestyle

33. WAS AN AUTOPSY PERFORMED? ☐ Yes ☐ No
34. WERE AUTOPSY FINDINGS AVAILABLE TO COMPLETE THE CAUSE OF DEATH? ☐ Yes ☐ No

To Be Completed By: MEDICAL CERTIFIER

35. DID TOBACCO USE CONTRIBUTE TO DEATH? ☐ Yes ☐ Probably ☒ No ☐ Unknown
36. IF FEMALE: ☐ Not pregnant within past year ☐ Pregnant at time of death ☐ Not pregnant, but pregnant within 42 days of death ☐ Not pregnant, but pregnant 43 days to 1 year before death ☐ Unknown if pregnant within the past year
37. MANNER OF DEATH: ☐ Natural ☒ Homicide ☐ Accident ☐ Pending Investigation ☐ Suicide ☐ Could not be determined

38. DATE OF INJURY (Mo/Day/Yr) (Spell Month)
39. TIME OF INJURY
40. PLACE OF INJURY (e.g., Decedent's home; construction site; restaurant; wooded area): Red Keep
41. INJURY AT WORK? ☐ Yes ☒ No

42. LOCATION OF INJURY: State: City or Town:
Street & Number: Apartment No.: Zip Code:

43. DESCRIBE HOW INJURY OCCURRED: Drank poisoned wine
44. IF TRANSPORTATION INJURY, SPECIFY: ☐ Driver/Operator ☐ Passenger ☐ Pedestrian ☐ Other (Specify)

45. CERTIFIER (Check only one):
☐ Certifying physician-To the best of my knowledge, death occurred due to the cause(s) and manner stated.
☐ Pronouncing & Certifying physician-To the best of my knowledge, death occurred at the time, date, and place, and due to the cause(s) and manner stated.
☐ Medical Examiner/Coroner-On the basis of examination, and/or investigation, in my opinion, death occurred at the time, date, and place, and due to the cause(s) and manner stated.
Signature of certifier: George R. Martin

46. NAME, ADDRESS, AND ZIP CODE OF PERSON COMPLETING CAUSE OF DEATH (Item 32): George R. Martin

47. TITLE OF CERTIFIER: Coroner
48. LICENSE NUMBER: 001
49. DATE CERTIFIED (Mo/Day/Yr): 4/18/1467
50. **FOR REGISTRAR ONLY**- DATE FILED (Mo/Day/Yr)

To Be Completed By: FUNERAL DIRECTOR

51. DECEDENT'S EDUCATION-Check the box that best describes the highest degree or level of school completed at the time of death.
☐ 8th grade or less
☐ 9th - 12th grade; no diploma
☒ High school graduate or GED completed
☐ Some college credit, but no degree
☐ Associate degree (e.g., AA, AS)
☐ Bachelor's degree (e.g., BA, AB, BS)
☐ Master's degree (e.g., MA, MS, MEng, MEd, MSW, MBA)
☐ Doctorate (e.g., PhD, EdD) or Professional degree (e.g., MD, DDS, DVM, LLB, JD)

52. DECEDENT OF HISPANIC ORIGIN? Check the box that best describes whether the decedent is Spanish/Hispanic/Latino. Check the "No" box if decedent is not Spanish/Hispanic/Latino.
☒ No, not Spanish/Hispanic/Latino
☐ Yes, Mexican, Mexican American, Chicano
☐ Yes, Puerto Rican
☐ Yes, Cuban
☐ Yes, other Spanish/Hispanic/Latino (Specify) ____

53. DECEDENT'S RACE (Check one or more races to indicate what the decedent considered himself or herself to be)
☒ White
☐ Black or African American
☐ American Indian or Alaska Native (Name of the enrolled or principal tribe) ____
☐ Asian Indian
☐ Chinese
☐ Filipino
☐ Japanese
☐ Korean
☐ Vietnamese
☐ Other Asian (Specify) ____
☐ Native Hawaiian
☐ Guamanian or Chamorro
☐ Samoan
☐ Other Pacific Islander (Specify) ____
☐ Other (Specify) ____

54. DECEDENT'S USUAL OCCUPATION (Indicate type of work done during most of working life. DO NOT USE RETIRED). King
55. KIND OF BUSINESS/INDUSTRY: Monarchy

Module 5: Teenage pregnancy answers

NCHS Definition of teenage births: Women ages 15-19.

Birth Certificate Assignment Instructor Answer Guide

This sheet provides an explanation and answer for the analysis of the Birth Certificates 1-51. The reasons are provided when necessary. Be sure you understand the responses as this type of problem will be on the test.

****Instructor's Note – Certificate 49 is a trick question which mentions a birth in 2012, which will not count towards a birth in 2011.**

Introduction: located in #1 (Includes Short Guide to knowing what to look for in the first certificate & rules to follow when analyzing the other certificates)

Looking at number 1

1. A case definition is a way to describe who has a health-related state by identifying what happened, where it happened and when it happened.

In this case you are interested in teenage births that occurred to residents of Tampa, Florida, during 2011.

Was the child was born in 2011? (This child was born in 2011).
Was the mother was a teenager (15-19 years old) when she gave birth? (Subtract the year of the mother's birth (1994) from the years of the child's birth (2011). This mother was 17 years old when she gave birth. The mother was a teenager when she gave birth to this child.)

Was this a Tampa, Florida, birth. (It is the location of the mother's residence, not the location of the hospital that determines where the child was born. Even though this mother gave birth in a hospital in Brandon, Florida, this would still be a Tampa, Florida, birth.)

Is this a Certificate of Live Birth for a baby born to a Tampa, Florida, teenager during 2011?
Answer **Yes.**

2. Is this a Certificate of Live Birth for a baby born to a Tampa, Florida, teenager during 2011?
   Answer **Yes.**

3. Is this a Certificate of Live Birth for a baby born to a Tampa, Florida, teenager during 2011?
   Answer **Yes.**

4. Is this a Certificate of Live Birth for a baby born to a Tampa, Florida, teenager during 2011?
   Answer **Yes.**

5. Is this a Certificate of Live Birth for a baby born to a Tampa, Florida, teenager during 2011?
   Answer **No**. Although, the child was born in 2011 and the mother was a resident of Tampa, Florida, the mother was 20 years old.

6. Is this a Certificate of Live Birth for a baby born to a Tampa, Florida, teenager during 2011?
   Answer **Yes.**

7. Is this a Certificate of Live Birth for a baby born to a Tampa, Florida, teenager during 2011?
   Answer **Yes.**

8. Is this a Certificate of Live Birth for a baby born to a Tampa, Florida, teenager during 2011?
   Answer **No.** Although the child was born in 2011, the mother was 12 years old, and she was a resident of Wesley Chapel, Florida.

9. Is this a Certificate of Live Birth for a baby born to a Tampa, Florida, teenager during 2011?
   Answer **Yes.**

10. Is this a Certificate of Live Birth for a baby born to a Tampa, Florida, teenager during 2011?
    Answer **Yes.**

11. Is this a Certificate of Live Birth for a baby born to a Tampa, Florida, teenager during 2011?
    Answer **Yes.** But this may cause some confusion. When you subtract the year of the mother's birth (1991) from the years of the child's birth 2011, you will find 20 years difference. However the mother was born in October 1981 and her child was born in September 2011, making her 19 years old when her child was born.

12. Is this a Certificate of Live Birth for a baby born to a Tampa, Florida, teenager during 2011?
    Answer **Yes.**

13. Is this a Certificate of Live Birth for a baby born to a Tampa, Florida, teenager during 2011?
    Answer **Yes.**

14. Is this a Certificate of Live Birth for a baby born to a Tampa, Florida, teenager during 2011?
    Answer **Yes.**

15. Is this a Certificate of Live Birth for a baby born to a Tampa, Florida, teenager during 2011?
    Answer **Yes.**

16. Is this a Certificate of Live Birth for a baby born to a Tampa, Florida, teenager during 2011?
    Answer **No.** Although the child was born in 2011 to a mother who was a resident of Tampa, Florida, this mother was 13 years old when her child was born. According to the National Center for Health Statistics' case definition, a live birth to a woman between 15 and 19 years old, this would not be considered a birth to a teenage mother.

17. Is this a Certificate of Live Birth for a baby born to a Tampa, Florida, teenager during 2011?
    Answer **Yes.** Although this may cause some confusion, the mother just turned 15, 10 days before the birth.

18. Is this a Certificate of Live Birth for a baby born to a Tampa, Florida, teenager during 2011?
   Answer **Yes.**

19. Is this a Certificate of Live Birth for a baby born to a Tampa, Florida, teenager during 2011?
   Answer **No.** Although the child was born in 2011 to a mother who was a resident of Tampa, Florida, this mother was 13 years old when her child was born. According to the National Center for Health Statistics' case definition, a live birth to a woman between 15 and 19 years old, this would not be considered a birth to a teenage mother.

20. Is this a Certificate of Live Birth for a baby born to a Tampa, Florida, teenager during 2011?
   Answer **No.** Although the child was born in 2011 and the mother was 17 years old, the mother was a resident of Brandon, Florida, not Tampa.

21. Is this a Certificate of Live Birth for a baby born to a Tampa, Florida, teenager during 2011? teenager during 2011.
   Answer **Yes.**

22. Is this a Certificate of Live Birth for a baby born to a Tampa, Florida, teenager during 2011?
   Answer **Yes.**

23. Is this a Certificate of Live Birth for a baby born to a Tampa, Florida, teenager during 2011?
   Answer **Yes.**

24. Is this a Certificate of Live Birth for a baby born to a Tampa, Florida, teenager during 2011?.
   Answer **No.** Although the child was born in 2011 and the mother was 19 years old, the mother was a resident of Cocoa Beach, Florida. Some confusion may occur about the mother's age. It can be shown that she doesn't turn 20 until 4 months after the baby is born.

25. Is this a Certificate of Live Birth for a baby born to a Tampa, Florida, teenager during 2011?
   Answer **No.** Although the child was born in 2011 and the mother was a resident of Tampa, Florida, the mother was 14 years old.

26. Is this a Certificate of Live Birth for a baby born to a Tampa, Florida, teenager during 2011?
   Answer **Yes.**

27. Is this a Certificate of Live Birth for a baby born to a Tampa, Florida, teenager during 2011?
   Answer **No.** Although the child was born in 2011 and the mother was a resident of Tampa, Florida, the mother was 20 years old.

28. Is this a Certificate of Live Birth for a baby born to a Tampa, Florida, teenager during 2011?
   Answer **Yes.**

29. Is this a Certificate of Live Birth for a baby born to a Tampa, Florida, teenager during 2011?
   Answer **Yes.**

30. Is this a Certificate of Live Birth for a baby born to a Tampa, Florida, teenager during 2011?
   Answer **Yes.** But this may cause some confusion. When students subtract the years of the mother's birth (1991) from the year of the child's birth (2011), they will find 20 years' difference. However the

mother was born in June 1991, and her child was born in January 2011, making her 19 years old when the child was born.

31. Is this a Certificate of Live Birth for a baby born to a Tampa, Florida, teenager during 2011? Answer **No.** Although the child was born in 2011 and the mother was a resident of Tampa, Florida, the mother was 13 years old.

32. Is this a Certificate of Live Birth for a baby born to a Tampa, Florida, teenager during 2011? Answer **No.** Although the child was born in 2011 and the mother was a resident of Tampa, Florida, the mother was 20 years old.

33. Is this a Certificate of Live Birth for a baby born to a Tampa, Florida, teenager during 2011? Answer **Yes.**

34. Is this a Certificate of Live Birth for a baby born to a Tampa, Florida, teenager during 2011? Answer **Yes.**

35. Is this a Certificate of Live Birth for a baby born to a Tampa, Florida, teenager during 2011? Answer **Yes.**

36. Is this a Certificate of Live Birth for a baby born to a Tampa, Florida, teenager during 2011? Answer **Yes.**

37. Is this a Certificate of Live Birth for a baby born to a Tampa, Florida, teenager during 2011? Answer **Yes.**

38. Is this a Certificate of Live Birth for a baby born to a Tampa, Florida, teenager during 2011? Answer **No.** Although the child was born in 2011 and the mother was a resident of Tampa, Florida, the mother was 20 years old.

39. Is this a Certificate of Live Birth for a baby born to a Tampa, Florida, teenager during 2011? Answer **Yes.**

40. Is this a Certificate of Live Birth for a baby born to a Tampa, Florida, teenager during 2011? Answer **Yes.**

41. Is this a Certificate of Live Birth for a baby born to a Tampa, Florida, teenager during 2011? Answer **Yes.**

42. Is this a Certificate of Live Birth for a baby born to a Tampa, Florida, teenager during 2011? Answer **Yes.**

43. Is this a Certificate of Live Birth for a baby born to a Tampa, Florida, teenager during 2011? Answer **No.** Although the child was born in 2011 and the mother was a resident of Tampa, Florida, the mother was 14 years old.

44. Is this a Certificate of Live Birth for a baby born to a Tampa, Florida, teenager during 2011?

Answer **No.** Although the child was born in 2011 and the mother was 17 years old, the mother was a resident of Orlando, Florida.

45. Is this a Certificate of Live Birth for a baby born to a Tampa, Florida, teenager during 2011? Answer **No.** Although the child was born in 2011 and the mother was a resident of Tampa, Florida, the mother was 14 years old.

46. Is this a Certificate of Live Birth for a baby born to a Tampa, Florida, teenager during 2011? Answer **No.** Although the child was born in 2011 and the mother was a resident of Tampa, Florida, the mother was 21 years old.

47. Is this a Certificate of Live Birth for a baby born to a Tampa, Florida, teenager during 2011? Answer **Yes.**

48. Is this a Certificate of Live Birth for a baby born to a Tampa, Florida, teenager during 2011? Answer Yes.

49. Is this a Certificate of Live Birth for a baby born to a Tampa, Florida, teenager during 2011? Answer **No.** Although almost the entire pregnancy took place in 2011, the child was born in 2012, right after midnight at 12:25am.

50. Is this a Certificate of Live Birth for a baby born to a Tampa, Florida, teenager during 2011? Answer **Yes.** Although, almost the entire pregnancy took place in 2000, the baby was born in 2011, right after midnight at 12:02am.

51. Is this a Certificate of Live Birth for a baby born to a Tampa, Florida, teenager during 2011? Answer **Yes.** Although the birth took place in Brandon, Florida, the child was born in 2011, to a mother who was 17 years old and who was a resident of Tampa, Florida.

Summary

Totals: 35 Yes
16 No

Module 5: Study design answers

1. Are individuals with a low intake of dietary iron more likely to **develop** anemia than individuals with a high intake of dietary iron?

   Cohort. This is a cohort study because **individuals are selected on exposure** (dietary intake of iron) are followed over time to determine if they will develop anemia in the future.

2. Among premenopausal women, is their dietary iron intake associated with their hemoglobin level?

   Cross Sectional This is a cross-sectional study because individuals are not selected on either exposure or outcome and **both exposure and outcome are measured at the same time.** It is a **snapshot** of health.

3. Is the mean *per capita* intake of dietary iron associated with the prevalence of anemia in a population?

   Ecologic This study compares results across populations and does not use individual measures.

4. Are teenage girls with anemia more likely to have had a history of a low intake of dietary iron than teenage girls without anemia?

   Case Control. This study identifies people by disease (anemia, yes-no) and looks back to identify exposure.

5. Among women with anemia, will intake of a dietary iron supplement result in normal hemoglobin levels?

   Clinical trial. This study is an **active experiment** that will give a treatment to people to see if it has an impact.

Module 6: Causality answers

Answers to 1-4

Table 1. Koch's postulates

1. The organism must be observed in every case of disease. Pedro was only able to find a positive culture in 75% of people with disease. Leads me to say "no".	No
2. The organism must be isolated and grown in pure culture. Pedro was able to grow the organism in culture. Leads me to say "yes".	Yes
3. The pure culture must, when inoculated into a susceptible animal, reproduce the disease. Pedro injected the bacteria from the culture into 15 male volunteers, 10 of whom developed the rash in 3 days. Leads me to say "yes".	Yes
4. The organism, must be observed in, and recovered from, the experimental animal. Sam next drew blood from all participants and Pedro set up cultures but none grew the gram positive bacteria. "Leads me to say "no"	No

5. How convincing is the evidence that the disease is due to an infectious agent?

   B. Somewhat convincing

   C. Not very convincing

Either somewhat convincing or not very convincing is correct as some of Koch's postulates were supported but some were not. There is room for uncertainly.

Consider the causality guidelines described above and identify which of these guidelines support causality by the caviar. For each guideline, choose Yes (Y), No (N), or Not Tested (NT).

Table 2. Association of mystery illness with eating caviar using the guidelines for causality.

6. Temporal relationship They did eat the caviar first	Yes
7. Strength of the association The odds ratio was low and not significant	No
8. Dose-response People who ate more did not get sicker	No
9. Replication of findings This theory was not evaluated in other studies	Not tested
10. Biologic plausibility The evidence from the rats showed a biologic possibility	Yes
11. Consideration of alternative explanations No other alternatives were considered	Not tested
12. Cessation of exposure It was not possible as people kept eating the caviar	Not tested
13. Specificity of exposure Other symptoms were caused by the caviar, such as diarrhea	No
14. Consistency with other knowledge No other studies available to compare to	Not tested

15. How convincing is the evidence that the disease is due to eating caviar?

B. Somewhat convincing
C. Not very convincing

Again, some of the guidelines were met but some were not. The most concerning is the lack of evidence to support the strength of association as evidenced in the low odds ratio which was not significant.

Consider the causality guidelines described above and identify which of these guidelines support causality by drinking Oval Office water. For each guideline, choose Yes (Y), No (N), or Not Tested (NT).

Table 3. Association of mystery illness with drinking water using the guidelines for causality

16. Temporal relationship All drank water before becoming ill	Yes
17. Strength of the association The odds ratio was low and not significant	No
18. Dose-response There was no measure of length or frequency of exposure	Not tested
19. Replication of findings This theory was not evaluated in other studies	Not tested
20. Biologic plausibility This was not evaluated as there were no data available	No or not tested
21. Consideration of alternative explanations Nothing was considered	Not tested
22. Cessation of exposure This was not evaluated	Not tested
23. Specificity of exposure No other diseases were reported by people who drank Oval Office water	Yes
24. Consistency with other knowledge This was not evaluated	Not tested

25. How convincing is the evidence that the disease is due to drinking water from the Oval Office?

B. Somewhat convincing
C. Not very convincing
D. Not convincing at all.

Again, some of the guidelines were met but some were not. The most concerning is the lack of evidence to support the strength of association as evidenced in the low odds ratio which was not significant. Either C or D are correct but some may choose B. A is definitely not supported by data. This is a question for you to think out the responses and not all epidemiologists would agree. The overall evidence does seem a bit weak.

26. Compile your results from the studies into the following table.

Table 4. Association of both risk factors with the mystery illness.

	Has mystery illness		Does not have mystery illness		Odds ratio
Exposure	Yes	No	Yes	No	
Ate caviar	20	20	50	60	1.2, ns
Drank water	25	15	61	49	1.3 ns
Ate caviar and drank water	23	17	30	80	3.6, sig

27. Using the concept of necessary and sufficient causes of disease described in number 5 previously, which of the four models best describes your results for eating caviar in Table 4?

B. Necessary but not sufficient

Individuals who both ate the caviar and drank the water were significantly more likely to become ill than those who did not do both activities, with a stronger measure of association (Odds ratio =3.6). But individually each was not a significant cause. Thus caviar was needed but water was also needed.

28. Using the concept of necessary and sufficient causes of disease described in number 5 previously, which of the four models best describes your results for drinking Oval Office water in Table 4?

B. Necessary but not sufficient

Individuals who both ate the caviar and drank the water were significantly more likely to become ill than those who did not do both activities, with a stronger measure of association (Odds ratio =3.6). But individually each was not a significant cause. Thus water was needed but caviar was also needed.

29. What public health warning will you give to the population?

C. Avoid eating caviar and drinking water"

While the odds ratios for each exposure was weak the two combined exposures were much stronger. The odds ratio for both exposures was 3.6, and it was statistically significant.

Module 6: "Brain Boosting" answers

1. Based on your reading in the background, which of the following would be your exposure?
   C. Taking Sintix or a placebo

Since participants are randomly assigned to take Sintix or a placebo, this is the exposure.

2. You also need to identify your outcome. In a clinical trial you can have more than one outcome. Which of the following would not be considered an outcome in the proposed study?
   a. Better concentration
   b. Increased test scores
   c. Taking Sintix or a placebo
   d. Increased anxiety
   e. Increased nervousness
   f. Poor sleeping patterns

The correct answer is C as other than taking Sintix, all of the other items can be measured as an outcome.

3. There are several hypotheses to be tested in this study. Which of the following do you think is a well written research hypothesis for the proposed study?
   C. College students who use Sintix will have 50% higher scores on a standardized test than college students not receiving Sintix.

"C" is the best answer because both the exposure (taking Sintix) and the outcome (scoring on a standardized test) are clearly identified and can be measures.

Ethical issues

Select Y for those warnings you would include and N for those you would not include. You need to know the side effects of Sintix. (Hint: look for patient information inserts (available online) and see Table 1 in the attached appendix.)

Table 1. List of potential items to be considered for inclusion in the informed consent.

Item	
4. Sintix may cause excessive anxiety. This was stated in the package insert.	Y
5. Sintix may cause excessive sleepiness. No, actually Sintix causes people to stay awake.	N
6. Students will be given either Sintix or a placebo, and will not know which one they get. This is true of a clinical trial and participants need to be informed of this.	Y
7. Sintix impairs the ability of the users to engage in potentially hazardous activities such as operating machinery or vehicles. They may not be able to drive when using it. This was stated in the package insert.	Y
8. Sintix may be associated with birth defects if taken by pregnant women. This was stated in the package insert.	Y
9. Sintix may cause increased blood pressure. This was stated in the package insert.	Y
10. Sintix may cause a decrease in concentration. Actually, Sintix increases concentration.	N
11. Sintix may cause weight gain. Actually, Sintix is associated with weight loss.	N

Selecting study subjects

Please indicate which participants should be invited to participate in the research study. Circle "I" for included or "E" for excluded.

Table 2. List of potential inclusion and exclusion criteria for the research study.

12. Students diagnosed with ADHD and currently taking Sintix These students could not be randomized to the control group as they need this medication. Also their use of the medication is different that that proposed in the study.	E
13. Students taking at least 12 credits in the Fall semester This can be an inclusion criteria. The researcher can pick a different number of credit hours, but this seems reasonable.	I
14. Both male and female students Yes, this study should include both genders as there is no good reason to exclude women, but protections need to be included. (See #15 & #18)	I
15. Sexually active female students not using any contraception These students can be excluded as they may become pregnant and Sintix is associated with birth defects.	E
16. Any student aged 16-25. Students under the age of 18 would need parental consent. They can be included, but it makes the study more complex as it would be difficult to get parental consent for college students who may be living far from home.	E
17. Students with a history of drug abuse Because of the potential for abuse with ADHD drugs these students should be excluded.	E
18. Pregnant students These students can be excluded as Sintix is associated with birth defects.	E
19. Only students at least age 18 This is a good idea from a practical point of view as these students can sign informed consents without needed a parental consent.	I
20. Spanish speaking students These students can be included as long as the study materials can be provided in Spanish if needed. Given the subjects are university students, most would be able to speak English as well as Spanish.	I
21. Students who live away from campus and only take online courses This would be a difficult group to monitor as they are away from campus and it would be more practical to exclude them.	E
22. Students already using ADHD drugs to improve test scores. Since these students already use ADHD medication, they may continue to do so even independently of the assigned medication. This could result in misclassification for the placebo group or be dangerous for the intervention group.	E

Study Procedures

Place a number from 1(first activity) to 8 (last activity) in the column on the right side of Table 3.

Table 3. Study activities.

Activity	Order
23. Obtain written consent from all participants.	2
24. Give participating students a test to determine their scores on a standard test of mathematical formulas.	6
25. Inform students of which medication group they are in.	8
26. Provide students with a study guide on mathematical formulas.	5
27. Give a screening survey to identify inclusion and exclusion criteria	1
28. Provide students with an envelope containing pills and instructions on the time they should take them.	4
29. Calculate test results. Survey the students to identify if they thought the medication they took had any side effects as well as whether they thought if it improved their test taking ability.	7
30. Randomly assign students to the two study groups: Sintix and placebo	3

Comments: (1) Usually subjects are screened to see if they qualify for the study before the consent is administered unless there are potentially harmful activities used to identify eligible subjects, such as blood tests.(2) Informed consent must be obtained before beginning study activities. Since people may refuse to participate when completing the informed consent this should be done before they are randomized. And for non-blinded studies, there may be different refusal rates between people assigned to different groups. (3) Subjects are randomly assigned to groups. (4&5) Either of these can be done first or together. Basically you want to give students the medication as well as the study guide to use to prepare for the test before they take the test. (6) The test is given next and then (7) a survey can be administered to students to assess their impressions before they learn what group they were in. (8) You should inform students of the group they were in once the study is over. You should also inform them of the study results.

31. In this study, neither the students nor the study coordinator knew which medication they were assigned. What kind of blinding was used in this study?
    C. Double Blinding

This is double blinding because both the participants and the person collecting data did not know which group the students were in.

32. Using the random numbers generated in Table 4, how many of the first 10 subjects received the Sintix and how many received the placebo? Read across the Table to answer this question.

    D. 6 Sintix and 4 Placebo

33. What is the main reason study participants or individuals are randomly assigned to the study groups?

    C. To decrease the bias in how study subjects were assigned to groups

Random assignment prevents bias in the assignment of individuals to study groups.

At the end of a research study, the Principle Investigator will need to analyze the data and present the results of the study. This is done in a manner similar to the M&M assignment you completed earlier. Refer to that assignment if you are uncertain how to answer the next three questions.

34. In this **hypothetical** study Dr. Romano reported the following result, "This study found that taking Sintix resulted in higher scores on the math test, with an average score of 76 among students taking Sintix as compared to 68 among students not taking Sintix, with a p-value of 0.13." What should he conclude?

    C. There was no significant difference in test scores between those on Sintix and not on Sintix.

A p-value of >0.05 means that the difference found between the two groups could be due to chance. So even though there was a difference between the two groups, it was not significant.

35. Dr. Romano also reported this second result, "This study found that college students taking Sintix were 2 times more likely to report difficulty sleeping than students not taking Sintix, with a p-value of 0.02." What should he conclude?

    a. Students on Sintix had significantly greater difficulty sleeping.

This p-value was <0.05 so it means that the study result was not likely due to chance and could be considered a true study result.

36. Based on the **hypothetical** results of this study, what would one conclude from the study?

    C. Sintix should not be considered for use as a brain booster among college students; as it is not useful in improving study scores and has significant side effects.

Thus is the most logical conclusion based on # 35 and 36.

The frequency of ADHD use as a Brain Boost among college students

Dr. Pillsbury plans to enroll 2,000 newly admitted freshmen students living in dorms and 2,000 newly admitted freshmen students living off campus, and then interview them when they are first enrolled and twice every year until they graduate or drop out of school. She will conduct interviews in September and May.

37. What type of study is this?
    B. Cohort

This is a cohort study because participants were picked by exposure (living in a dorm or not and followed over time to identify the outcome.)

38. What is the exposure in this study?
    B. Living in a dorm or off-campus

The exposure is living in a dorm because Dr. Pillsbury hypothesized that different living situations would result in different ADHD drug use.

39. What is the outcome?
    C. Using ADHD drugs

The outcome is using ADHD drugs which will be measured over time.

40. Dr. Pillsbury decided to only include students who were at least 18 years of age as they were able to consent for themselves. She is not sure what to do about students who already used ADHD drugs. What would you advise her?
    B. Exclude them because they already have the outcome and in cohort studies, you need to start with people who are free of the outcome.

In a cohort study, you need to only include people without the outcome when you start so you can identify the temporal sequence.

41. Both studies found that students who took ADHD drugs had higher test scores. Can Dr. Pillsbury report that taking ADHD results in a higher GPA as Dr. Romano was able to state for the math score in the clinical trial?
    C. Both B and C

Both B and C are correct as this is an observational study so the researcher cannot know if characteristics of participants that are related to the outcome (GPA) differed between the two groups to start with. This is in contrast to the clinical trial in which the two groups (Sintix and placebo) were considered to be similar in baseline characteristics because they were randomly assigned.

42. Can Dr. Romano state that his study showed that 50% of college students used Sintix as that is how many students used the drug in his study?
    B. No, because he is only measuring the number of students who he gave the drug to and not the number who used the drug on their own.

Since Dr. Romano is only measuring what he gave students and in fact he excluded students already using these drugs, he cannot estimate how many student use ADHD drugs.

Module 7: Disease screening answers

1. Screening for disease is conducted so that a disease can be identified in the early phase and early treatment can be initiated to prevent subsequent mortality from the disease. Thinking back to the information on prevention, what type of prevention occurs when screening identifies a disease early in the process?
   B. Sec*ondary*

Look at the 2x2 table presented below and write the letter corresponding to the correct formula for each component of screening on the line to the right of the word.

	Disease		
Test Results	**Yes**	**No**	
Positive	a	b	a+b
Negative	c	d	c+d
	a+c	b+d	

A: a/(a+b)

C. a/(a+c)

B. d/(b+d)

D. d/(c+d)

2. Sensitivity ____C____
3. Specificity ____B____
4. Positive Predictive Value ____A____
5. Negative Predictive Value ____D____

Which letter in the table refers to the each of the following?

6. True Positives ____a____
7. False Positives ____b____
8. True Negatives ____d____
9. False Negatives ____c____

Table 1. Calculations for diabetes prevalence of 2%

	Disease		Totals
Test	Yes	No	
Positive	198	490	688
Negative	2	9,310	9,312
	200	9800	10,000

10. What is the positive predictive value for the low risk community?
    A. 29%

11. In the community group what would be the total cost of the additional medical screening for each person who tests positive?
    C. $34,400

12. What would be the cost for each for each positive case identified? $173.73
    Hint: This is the total cost of additional screening/number of true positives.

Table 2. Calculations for diabetes prevalence of 20%

	Disease		Totals
Test	Yes	No	
Positive	1980	400	2380
Negative	20	7600	7620
	2,000	8,000	10,000

13. What is the positive predictive value for the obese individuals?
    B. 83%

14. In the community group what would be the total cost of the screening each person who test positive?
    C. $119,000

15. What would be the cost for each for each positive case identified? $60.10

16. Given Mark Phillips concerns, what screening strategy would you recommend?
    A. Targeted screening of obese individuals

Module 8: Outbreak Investigation

Crossword puzzle answers

Answer Key

							1 Z							2 T			
						3 F	O	M	I	T	E			O			
							O							X			
						4 I	N	F	E	C	T	I	V	I	T	Y	
							O							N			
							S					5 P					
				6 I			7 E	N	8 V	I	R	O	N	M	E	N	T
				N			S		I			I					
				D					R			N					
				E					U			T					
				X					L			S					9 C
10 A	T	T	A	C	K	R	A	T	E			O		11 H			A
				A					N			U		E			R
				S			12 V	E	C	T	O	R		R			R
				E					E			C		D			I
											13 V	E	H	I	C	L	E
														M			R
														M			
														U			
											14 A	G	E	N	T		
														I			
											15 H	O	S	T			
														Y